TAXI FROM ANOTHER PLANET

Taxi from Another Planet

Conversations with Drivers about Life in the Universe

CHARLES S. COCKELL

HARVARD UNIVERSITY PRESS

CAMBRIDGE, MASSACHUSETTS · LONDON, ENGLAND · 2022

First printing

Library of Congress Cataloging-in-Publication Data

Names: Cockell, Charles, author.
Title: Taxi from another planet: conversations with drivers about
life in the universe / Charles S. Cockell.
Description: Cambridge, Massachusetts : Harvard University Press, 2022. |
Includes bibliographical references and index.
Identifiers: LCCN 2021059776 | ISBN 9780674271838 (cloth)
Subjects: LCSH: Extraterrestrial beings—Public opinion. | Life—Public
opinion. | Evolution—Public opinion. | Taxicab drivers—Attitudes.
Classification: LCC CB156 .C615 2022 | DDC 001.942—dc23/eng/20220208
LC record available at https://lccn.loc.gov/2021059776

Dedicated to all taxi drivers
in the known universe

✳

contents

*

preface

*

It's a mysterious, beguiling, and fascinating material, this stuff we call life. As someone who spends their professional life studying it, I often get into conversations in all sorts of places about what it all means and whether life could exist on other planets. Whether it's parties or plane journeys, the matter of whether we are alone in the universe and why this grand experiment even started here on Earth can stimulate the most serious and most entertaining conversations. And there is one group of people whom I have found to be particularly interesting to engage in these sorts of conversations: taxi drivers.

Each and every day, taxi drivers are exposed to the bountiful and colorful menagerie of humanity. They start conversations with, or are forcefully regaled by, people from every walk of life, people holding every opinion. Left-wing, right-wing, religious, atheist, conservative, liberal, vegan, or meat-eater. Taxi drivers are linked into the collective mind of our civilization in a way that very few of us are. They feel the pulse of human thought. Not many other people boast continuous day-to-day exposure to such a wealth of human experience and outlook.

I don't get out much, and that isn't a comment of self-derogation. I suspect that it applies to most of us. I'm an academic and I write scientific papers with people who broadly

share the same worldview as I do. I attend scientific conferences where people talk and think about things that interest
me. When I do speak to people outside the safe paddock in
which my fellow horses roam, they usually ask me about science, so we end up talking about what I'm familiar with
anyway. I suspect that people in the corporate world do much
the same. Even real estate agents. I bet they don't talk much
about aliens, and when they are at parties, I have a suspicion
that they end up advising people on property. It's fine to be
like this. None of us can hope to grasp the totality of human
knowledge. Life is short. It's sensible to focus on a small area
of wisdom, assimilate it well, and then attempt to contribute
something to our civilization through that conduit.

But, having said that, it's enlightening to find out what
others think about some of the great questions we face. For
example, are we alone in the universe? I don't believe that this
question has escaped the attention of many people, whether
real estate agents or scientists. It's more than a scientific
question, though. It's a version of a question we ask ourselves
in our daily lives: Am I alone, in a physical sense, or alone in
some point of view that I hold? Being alone is a deeply human
experience. It's only natural that we wonder as a species
whether we are alone in a cold, vast, endless universe.

When we ask whether alien life exists, related questions
are sure to follow. Why should I care about aliens? If there
are aliens, what happens if they turn up in my hometown? If
the aliens are just a collection of writhing, squirming bacteria, smaller than the human eye can discern, would that
make any difference with respect to how I should treat them?
Why are we spending my tax money trying to answer these

questions anyway? And forget the aliens, will I ever get to go into space myself? What do all these questions mean for *my* life?

It was a muggy day in 2016 when I found myself in a taxi traveling from London's King's Cross railway station to 10 Downing Street. This isn't a journey I take every day. I was privileged to be asked to a party on behalf of the British astronaut Tim Peake, hosted by the prime minister. Peake was returning to Earth after completing a six-month expedition aboard the International Space Station. Along the way to Downing Street, the inquisitive cabbie asked me, "Are there alien taxi drivers?" That was when this book was born.

His question was the result of a chat that I frequently have with taxi drivers. It starts with an inquiry about where I'm going and why, meanders through a conversation about life in extreme environments, and ends with a foray into alien life and whether the apparently unlimited adaptiveness of life on our planet implies that the universe must be full of creatures. But while I've had this journey many times—it goes with the job—its outcome is always different. Like a random drive across the countryside, heading down side lanes and muddy roads, the conversation takes unlikely and surprising diversions.

When I give public talks about life in the universe, the format is always the same. I do my best to entertain the audience with interesting angles on this subject and then at the end, provided I haven't overstayed my welcome, they ask questions. It's different with taxi drivers. They don't have to wait for a presentation. As soon as you open that door and take your seat, they start the questioning, drawing on what they perceive to be important and probing what you have to say in response.

There is one thing that is common to all these discussions: they are always deeply interesting. Unencumbered by a cartload of academic knowledge, technical detail, and the conservatism bred by uncertainty, taxi drivers have clear perspectives on the sorts of questions that most people find significant. Sometimes, they offer an entirely new point of view. My journey that day in 2016 was a case in point. Name a single academic who would stand in front of 200 university students and ask, as though it were a profound question, whether there are alien taxi drivers. Yet here we were.

The question my driver asked was typical of many that this book contains. Seemingly simple questions often hide much more interesting ones; sometimes we can't answer them at all. For alien taxi drivers to exist, we need life to originate on a planet, we need those creatures to be intelligent, and we need them to invent economies and taxis. But how do you get from a few chemical compounds on a newly coalesced incandescent planet to an occupant of a taxicab? How many steps are there along the way, and how likely is it that one really will follow another? Where simple organisms exist, are intelligent organisms and complex societies sure to follow? In a flash of unencumbered thought, my driver had opened a Pandora's box of ideas about the possibility of life elsewhere and about the nature of our own society. Much that seems inevitable in life—biologically and culturally—starts to look contingent when observed through the prism of the alien question. Later that afternoon, as I stood wineglass in hand listening to Prime Minister Theresa May welcoming back Tim

Peake, the content of her speech went in one ear and out the other. I was thinking about alien taxi drivers.

What other questions about aliens, space exploration, and the phenomenon of life in general might taxi drivers have? After that day, I began to use taxi journeys as an opportunity to ask, talk, and think about life in the universe.

In this book, I've collected essays on some of the stimulating subjects that have arisen in my exchanges. I'll warn you now that all of these writings contain the unmistakable imprint of my own views. How could it be otherwise when each chapter emerges from personal conversations with taxi drivers? However, I have tried to give you some idea of where we stand in our knowledge and what the scientific community currently thinks about some of these questions. Some of these questions are about aliens—whether they exist, where they can be found, what they might be like. But the mystery of life in the universe is a many-sided one. What I hope this book will show you is that our curiosity about this subject speaks also to the scientific question of how life began, to political questions about whether we should explore space, and to profound issues of meaning in our own lives. I hope you'll jump in as I drive you on a tour across this landscape.

Perhaps somewhere, in a distant galaxy, there are alien scientists who write books about what they learned from their alien taxi drivers. Have many books like this been written across the known universe? Is this the first or the fiftieth? I don't know. Ask a taxi driver.

On planet Earth, the taxi driver is a ubiquitous feature of civilization, as in these examples from London. But are taxi drivers a universal outcome of biological evolution?

1

Are There Alien Taxi Drivers?

A taxi ride from King's Cross railway station to Westminster, to attend a reception for Tim Peake after his return to Earth from the International Space Station.

✳

It was a hot, stuffy day, and the underground was uninviting. I needed to get to 10 Downing Street without being late and, faced with a crowd of commuters, I left the station and hailed a taxi.

The driver, bespectacled and likely in his mid-forties, cheerfully asked me where I was going. I gave him the address, and as it was the home of our prime minister, his interest was sparked. What, he wondered, was I doing there? I informed him that an astronaut, Tim Peake, had returned from space, and the prime minister was throwing him a welcome-home party. I was one of the lucky invitees. Inevitably, this led to a conversation about my job, my interest in space exploration, and my fascination with the possibility of life beyond Earth. But it's boring and rather self-absorbed to sit in the back of a taxi and ramble on about your life. I wondered what *he* thought about the possibility of life on other planets, like Mars.

"Do you think there might be anything there?" I asked.

"Life on Mars, mate, I'm really interested in that, but what about aliens from other places in the universe?" he asked, cryptically. Maybe he was fishing for bigger things, for advanced aliens.

"Do you think there are intelligent beings out there?" I enquired.

"I'd say there has to be," he said. "There are so many stars and galaxies, they must be there. It can't just be bacteria, there must be something like us."

It seemed that he was interested in the subject. He spoke of bacteria and galaxies in the same sentence, fluidity that suggested that he had thought about these sorts of things before.

"What do you think they'd look like? Would they be like us?" I quizzed him.

"Well I think they would. One question I'd ask is . . ." there was a brief silence. Then, in an energetic and purposeful tone, he asked, "Are there any taxi drivers out there?" He paused again. "Are there taxi drivers like me driving around on other planets talking to aliens just like we're talking?" Another pause. "Yes, let me ask you this. Are there alien taxi drivers? People like me elsewhere in the universe?"

I've been a scientist for about thirty years, at least professionally speaking, and I've been to countless meetings, conferences, and workshops. In that time, I've heard a never-ending queue of fellow scientists discussing alien life. But on a brief taxi ride from King's Cross to 10 Downing Street, I heard one of the most cogent questions I have ever been asked: Are there taxi drivers on other planets? I couldn't let my driver down. He had asked a good question. So I'll tell you, in a little more detail, what I told him.

A taxi driver is a remarkable thing. Next time you spy one, you might like to contemplate how they came to exist. To make one, the churning, swirling matter of the universe must do several things in succession. To contemplate those steps is to understand why a taxi driver is really something to behold, and it is to delve into the question of whether taxi drivers are universal.

To start with, there is of course the question of how the universe came to be and why the universe is a taxi driver–friendly place. Are there other universes, parallel universes, where taxi drivers are forbidden by the laws of physics, where small tweaks in the fundamental constants that undergird the existence of matter deny the possibility of a cab? That's a matter for cosmologists. I'll pass on it and focus on our universe, whose physical rules permit taxi drivers. (That I evade this discussion is an extraordinary thing in itself, a telling reflection of how hard it is to explain the bare fact of existence, let alone the existence of taxi drivers.)

When the universe first formed, the rudiments from which it was constructed—hydrogen, helium, and plenty of radiation—were insufficient to build a taxi driver. This was true across the whole universe; everywhere was a pre–taxi driver cosmology. Indeed, taxi drivers, like all life on Earth, require at least six elements as a core part of their biochemistry: carbon, hydrogen, nitrogen, oxygen, phosphorus, and sulfur, sometimes called the CHNOPS elements. Apart from hydrogen, the other elements in this family were made in the cores of massive stars, objects in which temperatures and chemical reactions are so extreme that elements much heavier than hydrogen and helium could be fabricated. As these stars exploded, shedding the ingredients for taxi drivers across the universe, so these immense explosions would also call forth even heavier elements such as copper, zinc, and others that we find scattered throughout the biochemistry of taxi drivers.

This smorgasbord of elements now needed to be assembled into molecules that could replicate—the first inklings of living things. Otherwise, they would remain a collection of atoms swirling and mixing across the universe, and nothing more. How did a collection of atoms come together into those first molecules that began to replicate, to make many copies of themselves, but within which tiny variations would allow for improvements, for evolution? Despite decades of work, the answer remains a mystery. We still don't know how that first self-replicating chemistry took place over 3.5 billion years ago, leading eventually to taxi drivers and everything else we recognize in the living world.

It's not that we are completely at a loss to explain the transition from mere chemistry to biology. We know some of the basics: we need an environment that provides energy and the right chemical conditions to allow for the propitious chemical reactions that make a cell. There is no shortage of possible locations on our early planet where environments may have provided these Goldilocks conditions, just right for the emergence of life. From the belching vents at the bottom of the ocean that spew forth high-temperature fluids, to the interiors of ancient craters carved by asteroid and comet collisions, there is a cornucopia of spots where life-bearing reactions might have occurred. The exact ingredients in the recipe for life are a matter of debate, but we know that these ingredients were fabricated on Earth itself and in the swirling gases of the solar system. We find these same building blocks of life emerging in lab experiments that replicate conditions on the early planet and in meteorites, the remnant rocks of the dawn of our cosmic neighborhood.

But what first emerged from the brew of energy and chemicals is unclear. We don't know how those simple chemicals assembled into the metabolic pathways and replicating chains of a cell. That may have been a fluke, or it may have been inevitable. Here is our first bottleneck. If the trillions and trillions of chemical reactions occurring on a warm, wet planet inevitably lead to self-replicating, evolving biology—to life—then we are already nearer to our quarry: taxi drivers. But if this transition was one in an unfathomably large number—a chance event of such infinitesimal likelihood that it could not have repeated many times across even the vast universe—then taxi drivers are extraordinarily rare.

Once Earth's early replicating molecules came into existence, their journey to complexity was underway. One of their first achievements was to become encapsulated within a membrane—to become a cellular structure. Within the confines of their walls, the molecules could explore metabolisms and chemical pathways that would eventually allow them to adapt to the various environments found across our planet. New pathways allowed them to eat sulfur and iron as a source of food. Later, perhaps much later, sugars produced inside the cell may have helped some microbes survive desiccation on the early land masses. Over the course of a billion years or more, these cells, these microbes, would spread across the planet, exploring a vast diversity and combination of evolutionary possibilities as they worked their way into nooks and crannies from the polar ice caps to the searing interiors of volcanic pools. These early chemicals had, in essence, escaped the tendency of the world's oceans to dilute, disperse, and dissipate them. Cells conquered the world.

After these events and onward to the present day, the oceans and land masses teemed with microbial life. Today we think there are not a billion or a trillion of these living creatures, but a one followed by thirty zeros. There is no formal name for that number of things; it is too great. But microbes are limited in their complexity. The energy sources they use—hydrogen, ammonia, iron, sulfur, and others—can only allow them to build so much. An energy revolution was needed for those single-celled creatures to transmogrify into more complex forms that would ultimately become taxi drivers.

Long before the microbes had clocked up their billion-year birthday on Earth, quietly in the background, that revolution was already underway. The key is that some of the cells, known as cyanobacteria, developed the capacity to use sunlight and water as forms of energy. This new mode of gathering energy opened a vast empire, since any place that had these two ingredients could now be called home. This form of photosynthesis unshackled life from all those rocky minerals that so limited the places where cells could get energy; it allowed life to spread across the oceans and over the land.

The process of converting the sun's energy into the energy that powered cyanobacteria—and later algae, plants, and other photosynthetic organisms—involved new biochemical machinery that split water into hydrogen and oxygen. The hydrogen is essential to powering the cell, but the oxygen is a waste product, and the cyanobacteria belched it into the atmosphere. For a long time, this gas would make no real impact. Reacting with iron, hydrogen sulfide gas, and other gases in the primitive atmosphere, oxygen was mopped up and removed. But over time, those sources of oxygen-depleting reactions would come to an end, and the oxygen would begin to accumulate in the atmosphere, a consequence of the sheer abundance of photosynthetic life. It has sometimes been said that the cyanobacteria were responsible for one of the most enormous acts of atmospheric pollution of all time, but their microbial insouciance should not attract dismay, for the poor little things hardly knew what they were doing.

For some of those microbes that had hitherto lived happily in their oxygen-free world, the build-up of this new pollutant probably spelled disaster. Although we associate the gas with life, it is a chemically reactive substance and produces a kaleidoscope of reactive oxygen atoms and molecules, which would have attacked the unprepared by damaging crucial molecules such as proteins and DNA. Life exposed to oxygen would have needed to evolve defense mechanisms to protect itself from this onslaught. But every cloud of oxygen has a silver lining. When oxygen is combined with organic material—that is, carbon-rich molecules—the reaction can release great quantities of energy. Enter onto the stage aerobic respiration, the energy-gathering reaction used by you, me, and taxi drivers and seen in the less controlled display of a forest fire, when carbon-replete trees burn in oxygen.

With the breath of oxygen, life now had access to much greater quantities of energy and, with it, the possibility of cells cooperating and building animals. About 540 million years ago, oxygen, now composing a prodigious 10 percent or so of the atmosphere, enabled the emergence of animal life. Then, over time, the animals grew in size in a kind of arms race between predator and prey: larger animals could hunt more effectively, but also more readily escape being eaten. Oxygen initiated this process of experimentation in biological form.

The transmutation of single cells into animals was a vital step on the way to taxi drivers. Like the origin of life itself, it may or may not have been inevitable. Will life on any planet discover photosynthesis and then disgorge oxygen into the atmosphere? And even when that gas fills the sky, will life always use it to transmute into complex aggregates of living things, creatures that can run, jump, and fly? Can we imagine worlds where microbes alone smear the surface; worlds destined to end their days having seen nothing more than the slime of microscopic creatures? Here is another roadblock on the path from the basic building blocks of life to taxi drivers.

On our own blue marble, that transition did occur, and for hundreds of millions of years multicellular creatures blossomed and diversified into the biosphere we know today. Don't be too impressed though. Even now, the great majority of species on our planet are microbial; we live on a microbe world. Plants and animals are a late-coming phenomenon, dependent to this day on the microbial churning of elements that sustains their nutrient demands.

When I reached the end of this potted history of the emergence of life, my driver seemed surprised at the sheer number of events that occurred in those long stretches of time. He scratched his head and rolled down his window to get some fresh air. A warm blast hit my face. "So many things have happened to get here?" All that long-lost family history. I continued to bring him nearer to the answer he sought.

The animals began their march, I explained, but in no preordained or obvious direction. The dinosaurs gained mastery of the land, sea, and air for 165 million years. Yet in an instant, an object from space ended their evolutionary career and consigned them to the same fate as 99 percent of all animals that have ever lived: extinction. Over the course of ages, animals and plants continued unconsciously to diversify, blindly conforming to the laws of physics and following the path of evolutionary experimentation.

But then, about 100,000 years ago, an animal developed new and advanced tool-building abilities, the capacity to explore and learn in ways not seen before. The brain of this animal would grow to a size enabling self-awareness. In a geological blink of an eye, this animal would begin to leave artifacts of its powerful mental capacities: paintings, shaped arrows, pottery, and eventually space stations. What was the biological switch that made the emergence of consciousness and intelligence possible? These characteristics were once thought to be something categorically different from all that came before them, but we now know that many animals, from crows to fish, possess rudimentary tool-building abilities and are capable of some kind of cognition. The human brain is not fundamentally different, a throw of the dice that produced intelligence. But is it an inevitable development? Here too, we must humbly face our ignorance. This question speaks to whether intelligence is rare or common in the universe, but we don't have good answers.

These apes used their minds to collaborate. Upon realizing the vast benefits to be gained from coordination, they created agriculture, animal husbandry, and industry. They birthed societies—first foraging and agrarian communities, later megalopolises that would house millions.

As human communities grew, they needed better ways of moving resources around. The answer, provided by the ingenuity of the human mind, was a wheel. Potter's wheels were first made around 3500 BCE in Mesopotamia, and within 300 years they became the basis of chariots. At roughly the same time, we think, ancient Egyptians were experimenting with spoked wheels. The oldest wooden wheel to be discovered was found in Ljubljana, Slovenia, and is thought to date back to about 3200 BCE.

As the chariot and cart spread, an enterprising individual must have realized that spare cargo space could be used to transport a person to their preferred destination in return for some payment. With this single thought, the taxi driver was born. Given the appearance of the wheel by 3200 BCE, I put the appearance of the taxi driver shortly afterward; let's say about 3100 BCE.

In this supreme moment, when the first human being turned to another and said, "Yep, I'll take you to Jericho, mate, but it'll cost you a goat, and I expect a tip," taxi drivers came into existence on a planet orbiting an unremarkable star in the spiral arms of a galaxy drifting through the eternal emptiness of space. We might wonder whether this event, too, was inevitable. Are our commercial instincts an unavoidable result of the process of evolution? Can we imagine an alien civilization whose economics is based on altruistic cooperation, a society in which the idea of extracting payment in exchange for service would never occur? I suspect one could forcefully argue that, even in this speculative utopia, drivers might still ask for compensation to cover the basic upkeep of their vehicles. In any case, once communities of creatures gather and build complex societies, then transportation and vehicles, and thus taxi drivers, seem inevitable.

How far we have come. Over three and a half billion years ago, chemicals washing around on the surface of Earth transformed into replicating molecules, molecules that became enclosed in cells and harvested new forms of energy that eventually shapeshifted them into multicellular creatures. These life-forms evolved brains, emerged into self-awareness, invented wheels, and became taxi drivers. If the entire history of Earth were compressed into one hour, then the final stage in this epic, the period since the appearance of a taxi driver, would last about a five-hundredth of a second.

Throughout this saga, there have been branch points, episodes where life took a new direction: the emergence of replicating molecules, the formation of cells, the invention of photosynthesis, the appearance of animals and intelligence. We don't know if these changes had to happen and therefore if they have occurred all over the cosmos. If in fact any of these steps is unlikely, then our own planet may be a rare haven in a universe otherwise devoid of taxi drivers.

My own cab turned into Whitehall and pulled up outside the security gate to 10 Downing Street. As I came to the end of my taxi journey and this journey through time, my driver was sitting bolt upright, looking almost proud. It was as if, by contemplating his family tree from his grandparents back to the slime that covered the ancient Earth, he realized how special, how unusual he was. He grinned, we exchanged a fare and a thank you, parting ways.

Inevitable or not, it has taken very many microbes, a profusion of extinct animals, and mind-numbing amounts of time for our little world to make the voyage from mere atoms to taxi drivers. Every step along the way is captured in that one question—are there taxi drivers on other planets?

Next time you take a taxi ride, consider what a privilege it is to be a consciousness that can comprehend the spans of time and evolution that have given rise to the journey of life. Allow yourself to consider two amazing possibilities: that we live on the only world in the universe with taxi drivers, or that, out there, strewn across our galaxy and others, are many more taxi drivers, tentacled and talkative, driving their passengers across alien cities.

Historically, people took alien intelligence for granted. Thus in 1835 the *Sun* newspaper of New York was able to pull off an impressive hoax, convincing readers that, according to new observations, the Moon was home to winged humanoids and other animals.

2

Would Alien Contact Change Us All?

A taxi ride from Dulles Airport to NASA Goddard
Spaceflight Center.

✦

It was a crisp, cold evening in Washington, DC, and I was mildly jet-lagged. A long flight, followed by immigration control, waiting for a bag, and then a line through customs had left me in that transatlantic daze. I anticipated a bit of warmth and solace as I slid into the back of the taxi. As soon as I settled in, the driver proved keen to know exactly what brought me to town. He was in his fifties, I think, and larger than life, filling his seat with a rotund physique in a giant checked shirt that was fraying at the edges. He also filled the car with optimism; a face with a permanent grin.

"I'm here to talk to some colleagues about instruments for exploring other planets," I told him. "Going to the NASA Goddard Spaceflight Center." Now sometimes when I say this sort of thing, I get a sort of nodding affirmation and everything moves on. And sometimes I hit the jackpot—an alien enthusiast. On this evening, even though I wasn't in the mood for it, I had won the prize.

"So is there anything out there?" my driver asked in a no-nonsense manner. It's an interesting thing being an astrobiologist. People expect you to have an answer, to know something they don't. When you tell them that your guess is as good as theirs they seem perplexed, unimpressed even. So I asked my driver what he thought about the prospects.

"Well, it's frightening, isn't it? The aliens might cause disease, like in those movies. Maybe it would cause a disaster," he suggested with a convincing air of concern. His trepidation about the aliens was given extra richness by his almost melodic Southern drawl. Maybe Louisiana?

"But if they didn't cause disease, do you think people would care?" I asked.

"I wouldn't know, but if they were like us maybe they'd be helpful," he wondered.

"Would you try to contact them, or do you think we should just avoid them in case it all goes horribly wrong?" I asked.

"Well, they might give us their technology and then we'd get a lot out of it. That's it, you never know."

I wondered how he thought we might react, what the impact on human societies might be. "Do you think there would be chaos if we actually made contact with them?" I asked.

"If they came here, I think it would cause a great many problems," he said. "But if they sent, like you're saying, a signal. Maybe the media would have something to say. What am I going to do about that?" he questioned, his sentences short and to the point. He seemed genuinely disinterested in aliens that might turn up to Earth without something to offer.

I thought his reply was not atypical. Would aliens really change us? If you didn't have to deal with them directly, would they change your life? I nodded my head and agreed. My driver was quite indifferent to the idea of an intelligent alien civilization turning up on our doorstep. It is not an unreasonable response.

I wonder what you, reader, think about this. What do you think would happen to us if we found unambiguous evidence for the existence of an intelligent civilization beyond Earth? Would humanity erupt in a frenzy of discourse? Would our minds rise above our everyday concerns to confront the implications? Would we be terrified of contact, or might the experience bring us together at last, under the auspices of a new peace forged in the blinding alien light?

You might be surprised to learn that we know the answer to these questions, and we know them not merely in a speculative, conjectural way. We know the answers precisely.

In 1900 the French Academy of Sciences announced a new prize, the Prix Pierre Guzman, named for the son of one Anne Emilie Clara Goguet, whose bequest funded the award. In fact there would be two winners, splitting a hundred thousand francs. One prize was for achievement in medicine, the other for the first person to communicate with an alien civilization. But there was a catch. Mars was exempt because the prize committee figured that communicating with Martians would be too easy.

What gave the academy such confidence that life was out there? This perspective was certainly not new. The profundity of our place in the universe occurred to the ancient Greeks, leading to similar conclusions. Metrodorus of Chios, a student of Democritus who proposed a rudimentary atomic theory of matter, professed in the fourth century BCE, "It would be strange if a single ear of corn grew in a large plain, or there were only one world in the infinite." Of course, a

farmer invariably sows many seeds; that technical quibble
aside, Metrodorus was making the valid point that, where
conditions are ripe for life, there is usually a burgeoning of
living things, not just one. So too, Metrodorus reasoned, the
very fact that Earth exists should imply a plethora of earthlike worlds in the cosmos.

This logic—that life on Earth implies life elsewhere in the
universe—seems intuitively sensible. However, if there was
even one step in the origin of life that was mightily improbable, then Metrodorus might still be wrong; Earth might
be one living stalk in an otherwise-barren field. But he had
grasped with a beautifully powerful and simple piece of
thinking a question that resonated through the ages: Does life
on our planet imply life elsewhere? Metrodorus was one of
the first people known to have indulged the fascination with
the possibility of alien life, a possibility that would later grip
imaginations everywhere.

As the French Academy's rules demonstrate, Metrodorus's
optimism persisted. At the turn of the twentieth century,
it was widely thought that Mars was inhabited because it was
close to Earth and, being a rocky planet like ours, must also
harbor civilizations. Today such notions seem ridiculous,
not only because we know that Mars is devoid of alien societies but also because it is challenging to get into the minds of
people so utterly certain that aliens existed. Nowadays, we become excited by any discovery that suggests Mars may ever
have harbored conditions enabling life. But for the organizers
of the Prix Pierre Guzman, Martian life was banal.

In the prize's Martian exception lies the answer to the
questions raised by my taxi driver—whether the certain

reality of aliens would dramatically affect human societies. It is instructive to keep in mind that we have lived through a stage in our history where people were not only convinced of the existence of intelligent civilizations beyond Earth but even took their existence for granted. We know that, at the same time, wars continued as usual; humanity did not achieve harmony. We also know that "aliens" encouraged discourse, but it was confined to books, a few intellectuals, and maybe a few dinner parties. For most people, life was unaffected. The Martians had little to do with rent or food prices. Why should anyone care? To some readers, this mindset from the past may be discouraging, yet within it is a reassuring reflection of our civilization's capacity to cope with the possible trauma of alien contact.

We should grant a couple of caveats. For one thing, the enthusiasts of the last century had not actually made contact with aliens. In some respects, the silence reassured them the extraterrestrials were not interfering with us. No one was at risk. The reception of a real signal from a distant civilization might elicit a very different response, the nature of which would depend on the signal itself. A message transmitted long ago from a faraway place would register differently from a signal originating inside our solar system or from an object drifting at its edge. A nearby signal might induce goosebumps. But if the Prix Pierre Guzman organizers cannot give us a full picture of what humanity would be like when encountering the certain presence of aliens today, they do provide a sense of one way we could react.

Another lesson of the French Academy is that thoughts of alien worlds harboring life are not in any sense limited to our current scientific age. Not only did the possibility move the philosophers of ancient Athens, but also the Renaissance and the Enlightenment brought forward some astonishing ideas in this vein. One of the most startling speculations about worlds beyond Earth was proffered by the Dominican friar, mathematician, and philosopher Giordano Bruno. Born in Naples in 1548, Bruno traveled all over Europe, learning and writing. In 1584 he published a tome, which would not be out of place in a modern bookshop: *On the Infinite Universe and Worlds*. Buried within it is this arresting proposition:

> In space there are countless constellations, suns and planets; we see only the suns because they give light; the planets remain invisible, for they are small and dark. There are also numberless earths circling around their suns, no worse and no less than this globe of ours. For no reasonable mind can assume that heavenly bodies that may be far more magnificent than ours would not bear upon them creatures similar or even superior to those upon our human earth.

This was an impressive speculation on alien life for the sixteenth century. Not least, Bruno was talking about exoplanets more than four centuries before they were discovered. He had clear insight into why earthlike planets around distant stars are hard to find: because they are small and dark. Few of his contemporaries could even conceive of the idea that there might be something in space beyond what the eye could see or that brightness and dimness had anything to do with distance.

Sadly, Bruno was unable to follow up his ideas. He was arrested by the Inquisition even before his book was published, imprisoned for seven years, and burned at the stake in 1600 for various indiscretions vis-à-vis his church seniors and for harboring beliefs that irritated the Catholic hierarchy. It is thought that his endorsement of a so-called plurality of worlds—the idea that there are other earthlike planets in the universe providing homes for creatures—was one of the heresies for which he was charged. A multitude of worlds threatened the special place of humans in God's creation. It is sobering to reflect on the fact that, at one time, you could be incinerated for talking about exoplanets.

With the invention of the telescope in the seventeenth century, the late Bruno gained many fellows in speculation. One might reasonably assume that the opposite would have occurred: that the age of the fantastic would come to an end, replaced by an era of robust empirical observation. But it was not to be. Yes, humans now could actually see the other planets in the solar system, of which previously we had only hints. We could now verify the vast distances to the stars with greater accuracy, too. But if telescopes showed us that the maneuvering specks in our neighborhood were in fact planets, the resolution of these scopes was insufficient to make out surface details. So our ancestors had new planets to muse on but were no better placed to understand their life-limiting extremes. Speculation and fancy ran riot. This plethora of new worlds merely added to the number of potential alien homes, engendering the assumption that aliens were commonplace.

The solar system, it seemed, abounded with an infestation of societies.

The sheer bravado of alien speculation in the telescope era is sometimes hard for the modern mind to accept, particularly considering that many of the wildest ideas came from some of the most cogent and brilliant minds of the age. Christiaan Huygens, who invented the pendulum clock and discovered Saturn's moon Titan, wrote prolifically on extraterrestrial life and the habitability of other planets. In an elaborate synthesis of his writing on alien worlds, published in his posthumous 1698 book *Cosmotheoros,* Huygens speculated about astronomers on Venus and suggested that other intelligences would understand geometry. He appreciated that he had no evidence to support such statements, but this did not stop him. "This is a very bold assertion," Huygens wrote, "but it may be true for aught we know, and the inhabitants of the planets may possibly have a greater insight into the theory of music than has yet been discovered among us."

To today's reader, this is a highly enigmatic suggestion, but it can be better understood when we keep in mind that seventeenth- and eighteenth-century thinkers were often polymaths, facing no pressure to burrow into one narrow area of study, as contemporary academics do. Huygens was no exception; he was the son of a musician and music theorist himself.

At the same time, the era's political philosophers were beginning to wonder whether climate might be one of the principal factors shaping the nature of peoples. Under such epistemic conditions, to stare out into the night sky and see a planet like Venus was to invite speculations on the cultures that might emerge on a world hotter than Earth. Perhaps the alien minds were more active and thus their comprehension of music impressive? After all, as Montesquieu observed, "I have seen operas in England and Italy; they are the same plays with the same actors; but the same music produces such different effects in the people of the two nations that it seems inconceivable, the one so calm and the other so transported." The author of the *Spirit of the Laws*, which was to provide some inspiration for America's founding fathers, even described a bizarre experimental proof: he froze a sheep's tongue and noticed that the tiny hairs on it, which he surmised were responsible for taste, were retracting. This was evidence, he thought, for the effects of cold temperatures on the nerves and therefore on operatic performance. Venusians, no less than Italians and Englishmen, were presumed subject to the influences of their environment.

For my taxi driver, the significance of Huygens's musical projections lies in, again, the banality of his speculation. The presence of alien intelligences in the solar system—never mind on faraway planets—was so obvious that the question was not whether we should suppose they existed at all. That much was clear: people knew enough to feel sure that there was intelligent life elsewhere. The matter at hand was how well they understood and composed music.

Scientific confidence was reflected in expectations about aliens that found their way into literature. Science fiction and science have always danced around each other as if in a waltz, and never so much as in the arena of alien life. Similarly, the new genre of popular science writing, equally optimistic about the prospects for aliens, ignited a fury of thought and discussion across the drawing rooms of Europe. Popular writers propagated a spirit of certainty about the existence of extraterrestrial entities. Of the many tracts and pamphlets that touched on alien life, none was as widely read as *Conversations on the Plurality of Worlds,* written by Bernard Le Bovier de Fontenelle and published in 1686. This easily digested little book about the inhabitants of the Moon and other planetary bodies was gripping and delightful. It was a heady mix of science fiction and the emerging scientific consensus. The book features a narrator, Bernard, as he converses in a moonlit garden with a marquise keen to know about the workings of the solar system. The book is timeless, an utter joy to read even today. I recommend you add it to your list.

It is difficult to place the book's quality, but for me, part of it lies in the persuasive and humble argumentation of Bernard. He comments often about his lack of knowledge and his caution in overstepping the bounds of known astronomy, yet he leaves you with the impression that only a madman would deny that the Moon is inhabited by a civilization. Thrown into the pot is the delightful demeanor of the marquise, an intelligent girl full of astute, even moving, questions. It is easy to see how the book gripped the minds of Europeans innocent of our modern astronomical insights and led many to an ardent belief in the existence of life beyond Earth. Fontenelle solidified the view on the street that intelligent aliens lived on our doorstep.

A hundred years of discovery did nothing to blunt the imagination. Enter onto the stage William Herschel, another luminary, discoverer of Uranus and infrared radiation. His musings on astronomy must surely be authority itself. Yet here he is in the late eighteenth century, writing on the Lunarians, those inhabitants of the Moon: "By reflecting a little on this subject I am almost convinced that those numberless small Circuses we see on the Moon are the works of the Lunarians and may be called their Towns."

Herschel had seen perfect circular features on the Moon, which he, like everybody of his age, did not understand to be caused by asteroid and comet impacts on the lunar surface. There is a curious thing about impacts. All craters formed by asteroids or comets, except those arriving from the most oblique angles, leave almost perfectly circular scars. Herschel was a reasonable person and, as such, he was convinced that no natural geological process could possibly produce so many perfect circles. Their geometric regularity suggested a mind at work, the products of an intelligence.

We need not detain ourselves with long philosophical musings about science, but Herschel's observations and speculations are a sure warning from the past about the desire to believe in aliens. Any chink in the armor, any tiny unexplained geological perfection or phenomenon that does not immediately admit of a straightforward explanation, and in the aliens swoop, ready to claim the uncharted ground. Even the best of us can become beguiled.

Popular science writing persisted, hot on the tails of the scientists. *The Plurality of Habitable Worlds* was just one book in a prolific series penned by the French astronomer Camille Flammarion in the second half of the nineteenth century. As its leading title suggests, the book presumes life elsewhere. It goes on to detail how aliens would adapt to their environments, suggesting that we could predict what other life-forms might look like based on the homes they inhabit. Now the scientific speculations, even in the popular realm, were getting serious.

Newspapers are supposed to report on facts, but editors, witnessing the alacrity with which the public took to aliens, poured fuel on the fire. Purporting to reprint scientific observations published in an Edinburgh-based journal, New York's *Sun* newspaper pulled off an extravagant hoax with a multipart exposition laying out the discovery of winged people and beaverlike intelligences on the Moon. Claimed to be the fruits of work by astronomer John Herschel, son of the aforementioned William, this deception ran throughout August 1835 and earned the newspaper huge circulation numbers; for a moment, it was the most-read newspaper in the world. Other papers across the globe slavishly reproduced the remarkable findings, while poor Herschel himself was assailed with letters about his "discoveries." A hoax it may have been, but such a grand piece of fiction can only be pulled off if the collective mind is receptive.

It is noteworthy that in all this fervor, human society was not moved one iota to change its ways. No one thought to point out that perhaps the Lunarians would be unimpressed to the point of ignoring us once they saw our wars and pervasive poverty. It did not occur to anyone that a common spirit of intellectual advance and political fraternity, rising above conflicts of class and nation, might be appropriate for a civilization that was now part of an interplanetary fellowship. The obstinacy of human conduct is a tough thing to crack.

Enthusiasm for sure-to-be-real aliens did not abate, even into the twentieth century. As late as 1909, Percival Lowell, observer of the infamous Martian "canals," wrote in his book *Mars as the Abode of Life,* "Every opposition has added to the assurance that the canals are artificial; both by disclosing their peculiarities better and better and by removing generic doubts as to the planet's habitability." Lowell was convinced that a dying Martian civilization had built the canals to channel water from the polar ice caps to its cities in a last desperate attempt to save itself from desiccation. For Lowell, this was not science fiction, but others, such as H. G. Wells, saw the appeal of a good story. Wells took humanity's brooding concern about aliens and gave it expression in his now-iconic *War of the Worlds* (1898), which told of the arrival of Martians and their machines. These aliens burned Victorian England with their death rays before succumbing to microbial infection. That is the eternal self-reinforcing dance of science and science fiction for you. Each egged on the other until a frenzy of alien activity had taken hold of the minds of the public.

Within this long history is a lesson on how we might respond to a real signal, how we might direct our attention to intelligent alien life, without fundamentally changing our outlook. Perhaps the human condition is too self-centered. Even the inquisitive glare of the Lunarians didn't encourage us to grow up a bit.

These centuries of optimism, speculation, and presumption only came to an end in the late twentieth century, when the space age finally allowed us to send robotic emissaries to the planets to observe them close up. Then we could see with our own eyes the barren wastelands devoid of music-making Venusians, the Martian canals having neither locks nor towpaths, and the desolate sun-soaked craters of the Moon with nary a Lunarian in sight. The age of alien civilizations was over.

Yet in this demise of the Lunarians, there was an interesting inversion of the casual acceptance of alien civilizations that had gone before. Now we had to face up to the fact that all the civilizations we had taken for granted had vanished in the ephemeral wisps of our imaginations. Yet there were no lamentations. Disappointment for sure—who wouldn't have been enthralled by a snapshot of Neil Armstrong and Buzz Aldrin accosted by a Lunarian? What stories they would have told of their visit with the Moon's immigration officers and their alien sniffer dog. But, even though none of that happened, our civilization did not collectively drop into a nihilistic paralysis, an introspective silence at our newly revealed loneliness in the solar system. We just carried on as before, as if nothing had happened.

And while we have confirmed that ours is, at present, the only inhabited planet in our corner of the cosmos, we have not lost interest in the possibility that there is yet life out there. New discoveries have reinvigorated both the search and enthusiasm for alien life. What of the discovery of habitable conditions on Mars and oceans under the crusty ice-covered surfaces of moons orbiting Jupiter and Saturn? Rocky worlds orbiting other suns, some of those worlds perhaps like Earth, have added to this recrudescence of optimism about alien life. Never will we return to the heady days of the Lunarians, but we can seek out alien microbes in our own solar system and intelligences in planetary systems far away.

Conversation has returned to the consequences of communication with equal or superior minds, or even the discovery of a single lowly Martian bug. Workshops and conferences explore, in a more professional way than did the speculators of former days, the social and political implications of alien contact. Even the United Nations takes an interest in alien life. If we think all this very novel, it is because we have forgotten the centuries in which humans were convinced that space was home to multiple civilizations with which we might communicate.

The Lunarians we thought were there left a fairly insignificant imprint on our societies and ways of thinking; there was a flurry of books and ideas, but today they amuse more than inform. We might look upon this history sullenly, wondering why we never bettered ourselves in preparation for a meeting that might come eventually. However, the lack of any noticeable impact on the progress or behavior of humanity might instead be a cause for some relief, a suggestion that perhaps we don't need legions of politicians and social scientists to ready humanity for aliens.

If we should ever exchange words directly with alien entities, the extraterrestrials will find themselves at the receiving end of a species that once thought creatures built ramparts on the Moon. We might just be unimpressed with them. After a few florid months of media interest and some good literature, we might just shrug our shoulders and continue on our way. If the aliens visit us, and they climb into my taxi, they may find a driver more interested in being paid his fare than getting the latest Galactic Federation news from the Omega Quadrant. I hope they won't be disappointed.

Orson Welles met with reporters after the broadcast of his 1938 radio adaptation of H. G. Wells's *War of the Worlds* caused panic among listeners fearful of an alien attack.

3

Should I Be Worried about a Martian Invasion?

A taxi ride from Leicester railway station to Exploration Drive, home of the UK National Space Centre.

✳

We pulled out of the station car park and, to tell you the truth, I hadn't given the meeting ahead of me a great deal of deep thought yet. I was happy to do it; I was headed to the National Space Centre, a fine museum, to talk about astrobiology education. But the last few days had been busy. The taxi ride was my opportunity to do a little planning and so I wasn't much in the mood for politics. But sometimes politics is what you are given. No sooner had I told the driver my destination than he was off.

"I'm not having a go at you, mate, but space—it's for rich people, isn't it?" he demanded. "I'm not going to go there and nor are poor people, so what's the point?"

I tried to mollify him. "Rich people definitely do have the money to fly there, but it's not only for rich people," I said. "There are lots of good things that benefit all of us—rich and poor—in space, like orbiting satellites for mobile phones or predicting our weather. And as well as the everyday benefits from all those satellites, we might even find something incredible and unknown. Don't you think it's quite exciting to look for life out there?"

"I don't really mind if they find life, so long as it doesn't come here," he said.

I was puzzled. After a pause I asked, "What do you mean by that? Why wouldn't you want that life here?" My driver seemed irritated. Bald and burly in his blue coat, he hunched over the steering wheel, which he held with an unyielding grip.

"I mean really life is what you make it," he replied. "If they are like us they might come and fight us, in which case I'm against them, but if they don't, good luck to them. As long as they don't come to Leicester. Leicester's not a bad place really, right? Life's what you make of it; you end up in a flower bed anyway. I don't care about all that stuff in the rest of the world. Rich people can go to Mars and if the microbe things are there, then that's fine. I've been here all my life, mate, and as long as life's OK here, that's fine for me. Leicester's problem isn't Martians, but there's no room for much of anything else, and there certainly aren't enough jobs to go around as it is."

We scientists, allowed the extraordinary privilege of contemplating and experimenting, feel some chagrin in situations like this, when we are confronted by those who would cast the question of alien life, or other questions of pure science, in the mold of rich versus poor. Such is the gulf between our rarefied concerns and those of everyday people just getting along with their lives.

Funny that these same economic and political anxieties did not afflict people during the centuries when alien arrivals were considered plausible. Aristotle forcefully argued that the

Earth was special and no other creatures like us could exist out there in the cosmos, but, as we saw in the last chapter, others felt exactly the opposite. For a long time, even under the carapace of religious acceptance, many people believed that God was never idle. Nature abhors a vacuum as they say, and in His enlightened way, the entire universe should have been filled with intelligent beings, making the best use of the space available. For centuries then, it seemed quite obvious that the universe was brimming with life, yet intriguingly there does not seem to have been any speculation that aliens might visit Earth and take our jobs. I wonder why that is the case. I suspect it was because we lacked any sense of the engineering that could allow us to travel to the planets. If you have no basis for planning such feats yourself, you cannot imagine how others would do it. Your default assumption would be that we all remain bound to our respective orbs, staring out into the populated universe but never visiting the other planetary islands. The practicalities and details of how an alien intelligence might reach Earth remained quite outside our scientific vision and so the corollary was that there would be no concern about their immigration.

Our thoughts about alien creatures are often mixed in with a fear of the unknown, an insecurity about the "other." When we did finally, in the nineteenth century, imagine the possibility of journeys across the interplanetary void, this newfound sense of possibility was quickly hijacked by those who saw a great cataclysm. H. G. Wells's Martian machines launched the first war between entire worlds; his vision of destruction tapped into the widespread, perhaps instinctual, fear of strangers. With this in mind, it was actually a surprise to me that a taxi driver in Leicester would be more concerned about aliens' job-seeking intentions than about the possibility that they would obliterate the United Kingdom.

Whatever worries there might be in Leicester, it remains the case that we have not yet observed aliens. This despite the fact that, unlike our alien-enthusiast predecessors, we now know that there are likely many worlds out there that approximate in some way the Earth. The past three decades have witnessed rapid advances along this particular scientific frontier, as researchers have discovered a large number of planets orbiting other stars. These so-called exoplanets turn out to be a zoo of possible abodes. Not all of them, of course. Most are nothing like Earth and so seem poor candidates for life. Some are ten times the size of the gas giant Jupiter. Some hug their star tightly, orbiting every few days while baking in its rays. Some are rocky, a little like Earth, but likely covered in deep oceans. And then there are probably earthlike worlds.

If they are home to intelligences, we haven't heard from them. The mystery of the alien silence is a captivating one. This odd stillness in the air often goes under the heading of the Fermi paradox, named for the physicist Enrico Fermi. How can it be that in a vast universe that harbors so many planets, some of which must be older than Earth, we have seen no evidence of intelligent aliens? There are whole books dedicated to the Fermi paradox; a litany of scientists have queued up to proffer reasons for the aliens' elusiveness. Perhaps they observe us but do not wish to interfere. Perhaps they are here already, but we cannot recognize them. Maybe they are out there but can't travel the unfathomable distances separating themselves from us. There is also the possibility that life is rare, that worlds that beget galaxy-traveling minds are so sparse that the rest of the Milky Way may be devoid of them.

As we drove through suburban Leicester, I looked out the window and for a brief moment relived childhood glimpses of Martian machines standing over houses, the death rays of *The War of the Worlds* searching for their victims. Leicester as the epicenter of an alien invasion, the demonic plans of the gurgling tentacled creatures thwarted by insistent ranks of angry taxi drivers blocking the entrances to the local branch of Jobcentre, the UK employment-services agency. Well, why not?

"I don't think you should be worried about the aliens coming to Leicester," I offered in a reassuring tone. "The thing is, if they are here, but they are remaining secretive, then they don't seem very interested in Leicester, or at least they seem intent on keeping their interest quiet. If they are out there and they have a problem getting here, then we can probably safely assume that unless next year happens to be the lucky momentous occasion in which they arrive to take up residence here, they are unlikely to pose any immediate problem." After a brief pause, I added, "If they are quite rare across the cosmos, then we should probably be more concerned about our isolation and separation. I think loneliness in space is more likely to be Leicester's fate, not alien invasion."

There is, of course, a much more prosaic reason that my driver needn't worry. Even if aliens did arrive and make themselves known, would they be interested in our jobs? It seems implausible. If they can travel the vast spaces between the stars, then they are unlikely to need much from us. What would they do with the money they earned? Maybe buy some food? Presumably they brought their own sustenance with them, whatever it is. It isn't obvious that, even if they were hungry, our biosphere would provide them anything palatable. Most of us need some adjustment to the food of foreign countries; tucking into the biota of an alien world might not be a wise culinary decision. If the aliens' biochemistry were different from ours, then much of our food might be useless to them. Apart from food, the aliens might want help to repair their ship or some resource to power it, but I doubt they would line up for jobs in an effort to realize those needs. They would either ask for them or take them.

So I'd wager that my taxi driver needn't be too worried about Leicester's job market, unless of course there are already aliens among us. But, of course, there are not. The overwhelming problem with abductions, UFO sightings, and other supposed ephemera of alien visitation is the generally appalling quality of the evidence. Anecdotes and fuzzy films make good books and television programs, but there is no solid reason to believe any of these claims. After decades of UFO hunting, there is still not a single set of data that would get through a half-decent peer-reviewed scientific journal. No matter how optimistic you might be, that must tell you something. Still, there are those who would have us believe that aliens have visited Earth and interacted with our society and that "the government" knows it and hides it. Let me say with the greatest of respect that, while governments can undertake magnificent accomplishments, and while we know very well that they keep secrets, hiding aliens and their spaceships for years on end is a challenge in the face of which they would be singularly useless. Bureaucrats have their limits.

We have assuaged some of our fears about alien invaders, but need we be worried about smaller things? My taxi driver was nodding in encouragement at what I told him, so I changed my tack and introduced him to some microbiology. I felt his positive body language meant that he was ready for some bacterial conversation.

"I don't think we need a threat of human-sized aliens to know that smaller life-forms, the microbes, can wreak havoc in our society," I suggested.

"You're right there," he interjected. "All those hospital infections and new diseases are our big problem," he replied.

"Do you think we should be worried about alien versions of them?" I asked. "I mean instead of intelligent aliens, microbes causing havoc?"

"I'd say definitely yes," he replied, with a tone that matched the conviction of his words. "We don't want them here either and we need to keep ourselves safe from them. I'd be just as worried about them as the intelligent ones."

Destruction wrought by *Yersinia pestis*—the bacterium that caused the Black Death—and more recently coronaviruses and other pathogens remind us that our technological accomplishments won't necessarily protect us from the tiny creatures that have mastered Earth for over three and a half billion years. If we can have so little trust in the smallest life-forms that share the planet with us—organisms that in many ways are our evolutionary kith and kin—what fate might await us when the arrivals to Earth are not taxi-driving, job-seeking sentients but instead microbes? When H. G. Wells needed to end his Martian invasion, it was to microbes that he turned, germs that felled the Martians and with them their towering machines of decimation. Can we be so certain that a relationship with alien microbes would not result in the same outcome for us humans?

Now you, reader, might be forgiven for thinking that I am off on a brazen foray into unconstrained speculation. But unlike intelligent aliens in search of jobs, alien microbes really have grabbed the attention of space agencies. Serious people worry that Earth could become contaminated if, say, the little critters were inadvertently brought here in samples collected from one or another rock hurtling through space. This fertile area of activity goes under the enticing heading of "planetary protection." NASA has a planetary protection officer, and the European Space Agency has a Planetary Protection Working Group.

The original goal of planetary protection officers, and one that still occupies them, was to prevent our own contamination of other worlds. The concern here is not for the welfare of aliens but rather for scientific rigor and research efficiency. We don't want to spend billions of dollars looking for life on Mars only to discover life that we took there from Earth. Microbes that hitch a ride in a spacecraft and then end up in your life-detection machine—or spill out onto the surface of a planet and find their way into someone else's machine— would mean considerable wasted time and money. Planetary protection was conceived to minimize these sorts of problems. Today planetary protection is overseen by the international Committee on Space Research. The committee doesn't make laws, but it does produce regulations that space agencies abide by under common agreement.

Stopping microbes from getting to other worlds is no easy task. In the 1970s, when NASA sought to make sure its Mars Viking landers were not replete with bugs that might interfere with its life-detection instruments, they cooked the spacecraft, turkey-style, for forty hours at 111°C. Today the more sophisticated electronics in spacecraft make the challenge of preventing contamination even harder, but ingenious scientists can use a variety of methods to remove bugs. Cold plasma technology or toxic hydrogen peroxide can be used to kill microbes and clean surfaces, minimizing the "bioload" on spacecraft and thus the forward contamination, as it is known, of other worlds.

In recent years the concern about forward contamination has taken on a more ethical dimension. Scientists are trying not only to maximize the quality of their experiments but also to minimize the chances of compromising an alien biosphere. Although we know of no other biospheres in our solar system, we cannot at this point rule out the possibility that they do exist. We should therefore proceed cautiously, taking measures to prevent spreading our own life-forms across the solar system. It's generally thought to be bad manners, and embarrassing, to be the space agency that accidently destroyed an entire planet's ecology.

What seemed to concern my Leicester taxi driver—much like his concerns about losing jobs—was the flipside of this problem. The planetary protection community calls this backward contamination: the uncontrolled arrival of alien life on our planet. NASA has been thinking about this problem since the Apollo project, when agency scientists realized that the astronauts would be returning to Earth with rock samples that might contain microbes. These days it is instead robots that visit distant surfaces with the express purpose of collecting samples and bringing them home. A key goal of such research is to figure out whether there has ever been life in, on, or near those rocks, so the discovery of microbes or evidence of them would be incredibly exciting. At the moment, scientists are figuring out how to retrieve a sample of Mars to investigate whether that planet once harbored life. More samples from asteroids and comets are to be collected and returned to Earth in the coming decades. And sitting in our vaults are the fruits of previous labors, including samples of the Moon collected by rovers and astronauts, and cometary and asteroidal rubble returned by various space agencies.

None of the locations we've sampled are home to any life we are aware of, so researchers and agencies tend not to worry that Earth has been exposed to anything dangerous. Any nervousness that does exist is motivated by the precautionary principle: although it is extremely unlikely that our samples are life-bearing, we should be careful because the consequences of introducing an alien microbe into Earth's biosphere could be catastrophic. It is therefore not surprising that space agencies use ultraclean facilities for handling extraterrestrial samples and take great care to seal them away and ensure that nothing leaks into the outside world. If a researcher wants to study the samples, that project must be done in a facility designed for the purpose.

But what of the danger? Should you really be worried at all? Probably not. Keep in mind that humans have lived with disease-causing bacteria and viruses for as long as there have been humans. These pathogens evolved with people, developing hand-in-hand with us over the ages as our immune systems sought to keep up with their changing designs, holding them at bay. Your body is an exquisite machine designed to track down and destroy the countless foreign particles you ingest and breathe in every day. It takes a very special bacterium or virus to circumvent our immune systems and do mischief. The common-cold virus, which mutates every year to give us new coughs and ailments, is a testament to the endless battle between our bodies and the viruses that seek to outwit us. The fortress of the human immune system is the outcome of millions of years of evolution—good thing too, since if every virus or bacterium that came along had an easy time taking up residence in your body, your life would be very short. Because our biochemistry is so good at responding to these incessant challengers, it is likely that we would fend off any alien bacteria or other biological entities that made it back here from a distant planet. Your body would detect the foreign particle and probably destroy it. The probability of a microscopic alien life-form in a returned extraterrestrial sample unleashing a pandemic is very low indeed. Leicester can sleep sound.

Another scenario is less easy to dismiss. Imagine a hungry microbe living on the frozen permafrost of Mars, deprived of food and eking out a living in those desolate extremes. Now imagine that the microbe is collected by a spacecraft that veers off course while heading to Earth and crashes in the polar regions. On being released from the craft, the microbe finds itself in the Arctic, a place that is reassuringly cold and in which there is food aplenty. The conditions are more suitable for growth than those on Mars. Now the microbe can proliferate, potentially ousting earthly microbes from their home turf and establishing itself in our planetary ecosystems. Such a scenario is more probable than a disease outbreak because, in this case, the invader does not need an organism to host it—an organism that will try to keep it out or destroy it. Instead, the microbe merely requires an environment where it can settle and divide.

Yet even this doomsday vision should not keep us up at night. For one thing, the probability that we will come across alien microbes is itself very low. But say, for the sake of argument, that we did collect such a sample. In that case a spacecraft crashing in precisely the earthly location where those microbes would thrive, and in the process releasing the microbes without destroying them, is also extremely unlikely. Still, returning to our precautionary principle, the unlikeliness of these events does not release us from the responsibility of minimizing the chances that they will occur. No one wants to have to explain to my driver why some of Earth's ecosystems were destroyed by a careless space mission, and that is good enough reason to treat samples from space as dangerous until they have been properly and thoroughly investigated.

We arrived at my destination, but I could not be satisfied that my task was completed until I knew whether I had allayed my driver's concerns. "What are your thoughts about the Martians now?" I asked.

"They're still not welcome in Leicester," he replied gruffly, "but it doesn't sound like they are coming anyway."

Leicester's taxi drivers are under no threat from a Martian invasion and neither are you. But whether our solar system harbors any life, and whether distant worlds orbiting other stars host aliens, remains a fascinating question. These creatures, if they exist, may never impinge on the daily lives of people going about their business on Earth, but whether they exist at all is something that we might all contemplate. As we continue the search for them, we need not concern ourselves with whether they will take our jobs, but we should approach our quest with an open mind and a degree of caution that befits the exploration of unknown frontiers.

Are environmentalism and space exploration separate challenges, or are they
indivisibly linked? Here, US astronaut Tracy Caldwell Dyson contemplates
Earth from the International Space Station.

4

Should We Solve Problems on Earth before Exploring Space?

A taxi ride to Paddington station, en route to
Heathrow Airport for a flight to the United States.

＊

As we drove through the crowded streets of London, I peered through the window and watched the bustle of people stopping and starting. They dodged among cars, eyes full of concentration as they judged whether the great forward advance of auto traffic, bicycles, and motorcycles would halt long enough for them to make a dash. So much focus on the brief and simple task of getting to the other side.

It seemed like the taxi driver had read my mind. "It's madness out there," he offered, hissing at the sound of a shopping bag knocking his bumper.

"I know. Everyone caught in their worlds," I replied. "So many problems to solve, so little time."

"What brings you out here?" he inquired, in a strong Indian accent. My driver was young and alert, perhaps new to the job. He wore a fashionable button-down shirt, and his arm dangled from his window.

I explained that I was on my way to the airport. Along with a few others, I had been invited to a NASA workshop to talk about the search for life among the frigid moons of the outer solar system. We would discuss what was known about life in freezing wastelands on Earth and then consider the major challenges facing NASA in the search for life in other exceedingly cold places. If life were trapped deep in frozen oceans, how would we detect it? And if we could detect it, how would we collect the rock-solid ice samples and ferry them home across the void?

"Wow, I'm really into that stuff," my driver interjected. "Sometimes you see it on TV. You know, space exploration and all that. You can't help being interested in it. But there are so many problems down here. We have to solve those problems first."

I looked out of the window again. How far were the minds of these hurried and stressed pedestrians from the moons of Jupiter? A world away, for sure. My driver hooted as a car cut in front.

"You're right," I said. "We do have a lot of problems here on Earth. There's no doubt about it. But does that mean we shouldn't dream of space and even visit other places? Perhaps we might find answers to some profound questions about what we are all doing here," I suggested.

My driver didn't hesitate. "I'm with you on that," he replied. "I'm with you on that. We can't think about traffic all the time, and space—it takes our minds away, takes us outside everyday concerns doesn't it? Might it solve some of our problems too? Perhaps our woes on Earth could be solved by looking into space."

His question was a cogent one. Many people are inclined toward the line of thinking that my driver initially articulated— that we should solve our problems on Earth before exploring space. Some people even think that dealing with earthly problems, such as environmental devastation, and missions of space discovery are two opposed efforts, that one detracts from the other. On further reflection, my driver had seen something that I think is at the crux of the matter: going into space and looking after the Earth support each other.

People are right that we should care about our home planet—intensely. Hardly a day passes without another glaring example of how more than 7 billion human beings and their mass consumption stress the planet. The diameter of Earth is just 13,000 or so kilometers, and here we all are, packed onto the surface of this tiny ball of rock. It hasn't taken long at all, geologically speaking, to erect recalcitrant piles of plastic waste and pillage resources to the point where delicate habitats have become impoverished.

Even the atmosphere we breathe is a thin, fickle, and limited thing. It can be difficult to grasp just how effortlessly it can be altered. The bulk of Earth's atmosphere is a mere 10 kilometers thick. If you could travel upward into the sky in your car at a sedate 40 kilometers per hour, you would exit the greatest mass of the air in about fifteen minutes. You couldn't drive across Edinburgh or Manhattan in that time. Our atmosphere is a gossamer, gently draped across the surface of Earth. Once you have apprehended its tenuousness, it is relatively easy to understand how pumping gases into it might change its composition. The rise in carbon dioxide doesn't require a momentous exertion of human industry. A couple hundred years of pollution will do.

So it is not at all surprising that many people balk at the idea of spending resources trying to get into space, study it, maybe even build settlements on Mars or the Moon. In the age of climate crisis, can we really justify spending money on space exploration?

But while this sort of thinking is entirely understandable, it misses a key point: we learn a lot about Earth by exploring space. In fact, the science of climate change itself has been greatly enriched by the study of our planetary neighbors, Venus in particular. Venus is a cloud-enshrouded hell of a world, a mysterious and enigmatic abode. People once dreamed that it was covered in swamps and creatures adapted to the warmer climate of a place much nearer than Earth to the sun. But as speculation turned to science, it became apparent that Venus was much too hot to support life; nothing can survive on a surface now known to bake at over 450°C. Yet this temperature is still much hotter than Venus's proximity to the sun would suggest. What causes these blistering extremes? We only started to figure out the answer about sixty years ago: the explanation lies in Venus's atmosphere. Packed with carbon dioxide, the atmosphere traps heat from the sun and prevents it radiating back out into space, thus warming the surface of the planet to temperatures at which liquid water cannot exist. Venus is a greenhouse world, a free lesson to astronomers, biologists, climatologists, and humanity at large on what happens when you put a lot of carbon dioxide into a planetary atmosphere.

The industrial production of carbon dioxide on Earth will never generate the quantities of gases seen on Venus, but the mechanism by which our planet is heated by this greenhouse vapor is exactly the same. It was by observing our sister planet that we saw for the first time how the greenhouse effect can shape conditions on whole worlds, warming them far beyond the temperatures they would naturally settle into under the rays of the sun.

The lesson here, one we ought to take to heart, is that Earth is not an isolated little sphere, hung under a primordial dome. We exist in a bigger environment, the vastness of our solar system. Within that great theater our history was decided and our future will be ordained. By exploring that great expanse, we foster the knowledge we may need to save ourselves.

If space exploration is teaching us about how not to destroy our planet, it also can help us protect ourselves from spaceborne hazards. That is where our taxi-ride conversation went next: asteroids and the danger they might pose to us.

Periodically our planet is bombarded by the remnants of the turbulent days in which our solar system was formed. Pieces of rock buzz our planet like a swarm of bees. Some of these rocks, so-called near-Earth asteroids, have the potential to intersect with our planet's endless journey around the sun; if those rocks are big, they could bring devastation to our home. This is not merely a theoretical concern. The asteroid that struck the Earth 66 million years ago, ending the long reign of the dinosaurs, is but one spectacular example. Much smaller asteroids also leave their mark. In the desert outside Flagstaff, Arizona, you will find a great cavity in the ground, like a giant had taken an ice-cream scoop and carved a one-kilometer bowl in the sand. This is the result of a tiny rock fragment slamming into the Earth about 50,000 years ago. The blast wave would have extinguished life for many hundreds of kilometers around it, flattening trees and bursting all asunder.

There are places like this all over the Earth, though they can be inconspicuous. One, in the bushveld of South Africa, is now filled with a salty lake, its slopes festooned with a verdant shrubby veneer. To an observer, it is one of countless pretty sights on our home planet, yet it records an extraterrestrial agent of destruction. These impact events seem somehow alien, the craters they leave scars of a primitive time. But the asteroids aren't done with us. Events of the scale that created the bushveld and Flagstaff craters occur about every few thousand years. If a rock like the ones that landed ages ago in Arizona and South Africa struck a city today, millions would be instantly dead.

You can merely ignore these events, but that would be foolish. With the threat now as clear as a flash of light in a dinosaur's eye, we should wise up to our inextricable link to the cosmos and act. To predict how often we will be struck by asteroids and whether our own populations are threatened, we must map these rocks in space, and to do that we need telescopes. We can search from the ground, but space telescopes do a far better job. They sit silently, sentinels scanning the heavens, unencumbered by the atmospheric distortions that limit their earthbound cousins. Alongside the trajectory and velocity of space rocks, it is also useful to know what they are made from, so we can tell how much damage they might do: Will they break up as they hurtle through the atmosphere, or will they land intact? To answer these questions we can send spacecraft to examine rocks in situ or collect samples from their surfaces, for study back on Earth.

If we want our civilization to avoid an astronomically spectacular ending, then the foregoing will have shown you that space exploration is inescapable. We need to study rogue objects, and that requires a space program. And when we find ourselves on a collision course with a dangerous asteroid, we will need a good deal of clever engineering to escape the consequences. That is the purpose of NASA's DART (Double Asteroid Redirection Test) mission. Launched in 2021, the 500-kilogram DART spacecraft is designed to collide with Dimorphos, a small asteroid that orbits a larger space rock, Didymos. The hope is that the impact will cause Dimorphos to change its orbit around its mother rock, a small perturbation that can be measured from Earth, demonstrating the principles of a technology to knock an asteroid off course. One shouldn't underestimate this mission. It is the first time that a species on our planet, after over 3.5 billion years of evolution, has tested a technology with the express purpose of saving itself from extinction. Across the ages, through the distant mists of time, there are a billion dinosaurs urging us on.

After I had discussed some of these rather scary facts with my driver, he looked enthralled but nervy. Probably an asteroid collision wasn't on his mind when he set out that morning. But he seemed convinced. Then, he hit on precisely the point that I think we all find challenging, when it comes to the question of prioritizing humanity's limited resources.

"I get all that, but you know, when you are driving around London taking passengers this way and that, the last thing on your mind is asteroids, right?" Indeed, I had the temerity to spend my time pondering these things in the middle of the day, when other people were doing proper work. He was right, of course. We shouldn't spend all our lives wondering about space. We have other tasks to do, like shopping, cleaning the house, and yes, doing our jobs. But I think we should find time to contemplate our universe, because doing so helps us obtain a wider view of what we are. This is an abstract way to put the point, which is fitting because our relationship to the rest of the universe should be a source of wonder. But that relationship also involves very real implications for our future.

I wanted my driver to understand the inseverable link between our planet and the space we live in, so I turned to a phrase environmentalists coined in the 1970s: Spaceship Earth. There is, embedded in this term, a simple truth. Earth is a giant spaceship, racing around the sun at 30 kilometers per second as the sun itself orbits the galaxy at 200 kilometers per second, a tango of stars and worlds making their way around a supermassive black hole in the center of the Milky Way. There is a big difference between the spaceships we build and the spaceship we live on: Earth's never-ending tour of the sun may seem rather pointless, whereas we build spacefaring vehicles to do specific tasks. But our planet is nonetheless a spaceship. Here we are, enveloped in a dazzlingly large life-support system we call the biosphere. It's far more complicated than even the clanking and whizzing machinery that feeds oxygen to the occupants of the International Space Station and scrubs their carbon dioxide waste from the air. That is entirely fair; it took a few years to build the station, whereas Earth's biosphere is the product of eons of evolution.

Environmentalists are really space engineers, studying and tweaking the planetary life-support system and exhorting us all to take care of it. The people who designed the International Space Station are small-scale environmentalists, attempting to refine and adapt their more miniature life-support system for the few people who depend upon it to live in space. Environmentalists and space explorers are one and the same people—both attempting to ensure the sustainable and successful presence of humans in the cosmos. They just work on different scales.

Now this may seem like stretching the imagination, but it is an important point. It is not uncommon to find environmentalists criticizing space explorers as time- and money-wasters, dreaming of space conquest or building homes on Mars even as we have immediate problems to solve on Earth. I have also come across space explorers who regard environmentalists rather dismissively. The thinking tends to be that environmentalists have an important agenda but are also somewhat inward-looking, infatuated with the wounds of Mother Earth when we could be probing the limitless frontiers of space. If instead we could only see that we are all aiming at the same goal—successfully living in this borderless environment we call space—then environmentalism and space exploration would seamlessly dissolve into the same vision of the human future.

From a practical standpoint, when we think about a sustainable human future, we should consider what space has to offer. As my driver veered off Edgware Road toward Paddington, I turned to a favorite analogy that clarifies the point. Imagine that one day, while out shopping, you are, by some accident of timing, locked into a store that is closing for the summer. Finding no obvious escape, you hunker down in isolation. Eventually you begin to run out of food and other resources, forcing you to stretch out what you have left. But why would you continue to remain shuttered in the store, rummaging for crumbs and morsels, when a good kick of the backdoor could release you onto a high street?

In the same way, Earth is limited. It's not that we shouldn't attempt to be efficient, to minimize waste and reduce the stress we put on the biosphere we exploit. But to relegate the whole human future to this one planet, relying on it to supply all of our energetic and material needs forever, is to close our eyes to the infinite bounty the universe has to offer. Earth probably has just a few hundred years of easily accessible iron ore remaining, yet in the asteroid belt, a rock-filled doughnut between Mars and Jupiter, there is enough iron to last us millions of years, not to mention platinum and other elements that could supply our high-technology industries. Mining the stuff that goes into our phones and computers is enormously costly for the Earth and, too often, for the laborers who do the work, as well as their families and communities. Wouldn't it be wonderful if we could obtain these necessary materials elsewhere?

Of course, like raw materials on Earth, resources beyond Earth are not always easy to extract. That's one of the reasons that space explorers and environmentalists are equally interested in solar energy, recycling, and mining technologies. Those who care for the Earth and those who wish to venture into space meet at this nexus, where both are committed to finding ever more clever ways to access, use, and reuse what they have. Here, too, there is enormous potential for great minds to come together to solve common problems. Whether they look to Earth or to space, engineers concerned with efficient resource use can help develop the means by which we can successfully thrive wherever we set up home.

It is certainly true that the economics of this future are uncertain. Can metals be mined from the asteroid belt in a way that makes it worthwhile for a public or private actor to expend the time and effort to do so? This is a question without a solid answer, although, given that we already see an increasing number of private companies launching objects off Earth, it seems likely that in coming decades space-based enterprises will increasingly fall into the realm of the economically possible. But such details need not concern us too deeply right now. We should think about the wider vision and recognize that we do not need to continue consuming this tiny speck of rock forever, hoping that it will continue to supply the many needs of 7 billion people or more. The universe beckons us outward.

This vision comes with a dark edge. We could imagine gathering the plentiful metals of the solar system and bringing them back to Earth, only to fuel mass consumption and the accelerated destruction of the environment. That would not be a sensible outcome. Instead, we have to do some planning. We could instead locate some of our industry in space, removing it from the claustrophobic pollution-collecting confines of Earth. If we gather metals from the asteroid belt, why not process them there as well? This future would be energized by a vision of Earth as an oasis in space, a proverbial residential district for our species, where parks, lakes, oceans, and the air are protected from the worst of our industry. We can leave the noxious gases to the pollution-forgiving vacuum of space.

As I made my case, my driver nodded enthusiastically. "And I suppose the space explorers learn important things about other planets right here?" he asked. We were pulling into Paddington station, where I would catch a train to Heathrow. But first I had to answer his question, because in many ways it closes the circle. Understanding space helps us live well on Earth, but the work we do on Earth also helps us settle the rest of space. Much of what we learn in this joint enterprise of exploring space and caring for Earth is mutually beneficial. They are not two separate objectives.

Today scientists travel to a plenitude of environments on Earth to prepare for the exploration of space. These so-called analog environments give us insights into the nature of other worlds and whether we might find life on them. In the frozen wastes of Antarctica, scientists investigate how life copes with extreme cold and desiccation and what that might tell us about the possibility of life on Mars. The way water behaves, periodically flowing from dark depths and shaping the landscape of some of the most extreme environments in the Antarctic and the Arctic, informs us about the ancient geology of Mars, once a world of lakes and glaciers melting under the light of a distant sun. My own scientific interests, and those of many of my colleagues, have been motivated by an interest in exploring life at the limits of its existence.

Indeed, where do those limits lie? We learn more about the answers by going to places where most people would not want to holiday—earthly locales that are frigid cold, broiling hot, arid, seemingly dead. Yet even in these places our biosphere hangs on tenaciously, giving us glimpses of the extremes where we may find life elsewhere in the universe. This sort of research can put your mind in odd places. I won't deny that lush forests seem to me decadent; yet, in the Arctic, I lovingly spare a tiny green layer or blob of life the wrath of my rock hammer. I see this sort of life, precariously balanced against death, differently.

From the parched landscapes of Chile's Atacama Desert to the Rio Tinto in Spain, a river as corrosive as battery acid, scientists have commandeered countless environments in this effort to understand the potential workings of alien landscapes and, on reflection, to better understand Earth. Earth's extremes are not an outlier in the universe; some of these conditions overlap with those we will find on other planets and moons. From these common dominions, we learn something about the histories of worlds and how human activities might influence Earth's suitability as a haven for life.

Although my driver intuitively grasped the link between Earth and space, the separation of environmentalists from space explorers is easy enough to understand. Space exploration was born from the Cold War, a battle for strategic and ideological supremacy in which space was the ultimate high ground. Projects that ushered forth from this confrontation were imbued with the spirit of competition—the first human into space, the first team into space, the first human on the Moon, and so forth. The battle over space waged by the United States and the Soviet Union hardly had any link to the environment. By contrast, back on Earth, concerns with the effects of pesticides and, more widely, the growing recognition of our environmental impact were encouraging a planetary awakening that seemed far removed from superpower conflict. Indeed, the competition going on in space was seen as antithetical to the peace-loving cries of the modern environmental movement.

It is not true to say that environmentalism and space exploration have always been polar opposites. Many organizations involved in the exploration of Earth embraced space exploration as the next great frontier, and astronauts have lauded the experience of being in space as a way to see the fragility and smallness of Earth. But in the wider scheme of things, the two activities have remained apart, fostering antagonisms and provoking the conviction that we should be solving problems at home before exploring space.

But there is a different way to view the human future. We need not see ourselves at a fork in the road, whereby we care for Earth or we explore space. This binary outlook understands these activities as mutually exclusive, emphasizing zero-sum choices between the two. Yet, in truth, both are suffused with scientific and technological benefits for each other. What we learn by exploring space can aid our understanding of Earth. And many space-based endeavors, like the mapping of asteroids and exploitation of resources, have immediate, practical benefits, helping us protect Earth and all that lives here.

Spaceship Earth. Here we are, orbiting the sun. Instead of viewing the cosmos as a distant distraction from our worldly problems, we should wholeheartedly embrace our place in the universe at large. In seeing Earth as bound into the very structure and fate of the solar system and its other planetary occupants, we will learn to better care for our home planet and understand how we can use space to our advantage, to enhance our chances of success and to supply our needs. We must not lose any time in addressing the critical environmental challenges that humans and the rest of the biosphere face, but at the same time, we should mobilize our capacity for space exploration to improve our future. Given the urgency of the environmental crisis, it may seem that space exploration should recede—that it is an unaffordable luxury. In fact, the promise of exploration is heightened, not diminished, by demands back home. Environmentalism and space exploration are twins, and through them we have a chance of arriving at a future in which Earth is a sanctuary, cared for by a spacefaring civilization.

After paying my fare, I thanked my driver and disappeared into the throng of passengers heading for the trains. Probably the last thing on their minds was asteroids or Mars. Yet it seems to me that, slowly but surely, the sort of conversation I had just had with my driver—about Earth and space, propelled by environmentalism and the successes to be achieved from exploration—is becoming a bit more common. As we reach further for the stars, perhaps space settlement will seep into the public consciousness much as awareness of Earth's imperiled future has. Perhaps taxi drivers will soon enough find that chatting about environmentalism, space settlements, and asteroids is just part of the job.

Space travel is coming to the general public. The SpaceX Dragon capsule, shown docking with the International Space Station, can carry private astronauts.

5

Will I Go on a Trip to Mars?

A taxi ride from the University of Edinburgh to Waverley station, to catch a train to London.

*

"Off to the railway station?" my driver asked. "Going anywhere interesting?" She was in her forties, I thought, and sported a pair of red glasses and a bouffant hairdo in ginger. She liked tapping on the steering wheel and repositioning her glasses every now and then.

"I'm going to the Rutherford Appleton Laboratory in Didcot to talk about an experiment we want to send into space," I explained.

"Space? Did you say space?" she peered inquisitively into the mirror.

"Yes, just an idea, but we're thinking about how we might get it to fly to the space station in the next couple of years," I continued.

Then she asked a question I often hear: "But will you go yourself?" Many people seem to think there is a real possibility of an answer in the affirmative. I wish there were.

"Unfortunately not. It'll take a little more for me to be an astronaut," I told her. "Perhaps when commercial rocket companies can offer a cheap enough ride, it might be an everyday occurrence for people like me to go into space, but not this time. But would you go yourself?" I asked.

She looked into her mirror, which was now almost filled with her wide eyes, staring into the back of the cab. She shifted her glasses and tapped on the wheel again.

"I'd go," she said, without hesitation. "Count me in. My other half wouldn't be particularly impressed. The children have left, so they don't care. But, can you imagine, what a chance! I'd go. Not for all my life. I'd like to come back, but I'd go."

I wondered what made her so eager. "Adventure!" she exclaimed. "Even if I wasn't the first, can you imagine that? I like my job, don't get me wrong, but Edinburgh, well it can get you down doing the same thing every day. Space would be different. I'd go if I was given the chance," she said.

She wasn't the first person I'd met who would throw themselves at the chance for a trip into space. Actually, I've always been surprised, even amused, by it. Lots of people whom you would never expect to harbor a burning desire to be flung into space in fact wish to go. Innkeepers, bankers, shop clerks, prisoners—there is hardly a walk of life that doesn't contain aspiring galactic travelers.

I had a very personal experience of this in 1992, while I was completing my doctorate. I was sitting in a pub in Oxford, exuberantly expressing my interest in Mars to my fellow students. It happened to be two months before our national elections, and my drinking companions suggested that I stand for office on a platform promoting Martian travel. So I did. They agreed to join my shadow government. We drove up to Huntingdon political constituency the next day, where the prime minister at the time, John Major, was a member of parliament. We collected the mandatory ten signatures, paid the deposit, and the Forward to Mars party was born. We put loudspeakers on my car (a Mini) and turned it into an election battle bus, driving around the streets and asking the public for their votes. Of course, to do that you need a catchy slogan. "It's time for a change, a change of planet" did wonders in putting smiles on street-goers' faces, so we stuck with that. Then we went to work with our quite serious manifesto of proposing that Britain set a goal to build a Mars station and increase its involvement in Mars exploration. I spent every Saturday afternoon of the whirlwind campaign giving off-the-cuff lectures about Mars exploration on the streets of Huntingdon and dropped in everywhere from radio stations to hospitals to churches to spread the good word.

Then came election night. Standing in trepidation beside perennial satirical candidates Screaming Lord Sutch and Lord Buckethead—oh yes, and the incumbent member of parliament—I received the verdict of the people: ninety-one votes. I had come second to last, beating the Natural Law Party but narrowly missing out on a seat in parliament by just a few tens of thousands of votes. Actually, I found the ninety-one votes quite alarming, given that I knew no one in Huntingdon. I still have no idea who my supporters were. But in those two months, I did find out a lot about what everyday people think of space exploration. The public's enthusiasm extends beyond watching TV specials. Indeed, the fact that you can turn up to something as serious as a national election and persuade ninety-one people whom you don't know to vote in favor of joining you on Mars says something about the thrill that a vision of space travel encourages. So to find the same excitement in a taxi driver wasn't much of a surprise to me, yet I continue to take note of the widespread interest in space travel—not just as a project for robots and jet pilots but also for the rest of us.

"How long will I have to wait?" she asked me, wondering when her opportunity might arrive.

Like others of my age, I recall the early, heady days of space exploration. I remember being about eight years old and reading a book about the NASA Apollo program. It was the mid-1970s, and the Moon landings were not exactly in the dim and distant past. Neil Armstrong's words still rung in many people's ears; the promise of where these grand exploits would eventually take us elicited a feverish excitement. In the back of the book, which detailed Armstrong's and Buzz Aldrin's lunar exploits, were two pages devoted to the glimmering future of the 1980s. The pages were draped with images of Mars bases and spaceships that would take you on far-flung journeys to the outer solar system. These possibilities seemed a long way off but still close enough to be tangible. Do you want to know something genuinely depressing? At age eight, I really thought I would travel to Mars in the 1980s.

The era's futurists foresaw space journeys as routine, not just for the lucky few who had the "right stuff" to become astronauts. Princeton physicist Gerard O'Neill filled out an entire book, *The High Frontier: Human Colonies in Space* (1976), with fantastical designs of space settlements. One was a giant toroidal spaceship surrounded by glinting mirrors to capture the rays of the sun and rotating slowly to simulate Earth's gravity. Inside, ten thousand space colonists grew crops, tended to houses, and built roads. O'Neill's images confronted us with a strange vision of two towns on opposing sides of a vast cylindrical space, so that one's brethren appeared suspended in the sky above.

It is easy, several decades on, to be mightily disappointed. It seems that all we have done is build one space station after another just above Earth, orbiting the planet in endless circles and going nowhere. First there were Skylab and Salyut, then Mir, now the International Space Station. Yet we still seem so far from Mars and have made no advances toward Moon settlements. But if it is reasonable to be disappointed, keep in mind that we have learned a lot in the decades since the Apollo astronauts visited the Moon; this time has not been sterile of progress. For one thing, we have learned about what being in space does to us, knowledge that will be critical if regular people are ever to go to space in numbers.

Space takes its toll on the human body, and the more time one spends there, the heavier that toll becomes, as muscles deteriorate in low gravity. A few days on the Moon collecting rocks and hitting golf balls is nothing compared to weeks in orbit. Subjecting tourists without the fitness of astronauts to these conditions is, for the moment, unthinkable. Even extremely fit astronauts must maintain a rigorous exercise regimen to keep them in condition while in space. They strap themselves down to run on treadmills and pull weights for a couple of hours every day to prevent muscular atrophy and bone loss. This is another health challenge of being in space: without force applied to them—like the force caused by gravity—bones tend to thin out and wither away.

There is more unpleasantness where that came from. It turns out that, without gravity to pull your fluids down, liquid tends to collect in the upper body, causing one's face to become bloated. And then there is that feeling of constant disorientation. Nothing is up or down. That computer over there? It could be on the floor, but it could be on the ceiling, depending on what you see on the opposite wall and what your brain tells you to expect of a ceiling or floor. This sensory confusion sends you into a spin. Your balance is in turmoil and you feel sick.

All in all, you need some practice in order to linger in space. Before astronauts had tried it, we had a sense of the difficulty. But now we have a much more refined understanding, as a result of relatively long periods of time space explorers have spent on orbiting stations. Although space stations might seem mundane to those of us stuck on Earth waiting for our ticket to Mars, they have been enormous producers of knowledge. Much excellent science has been done on space stations, from studies of antibiotics to research on the spread of fire. And, alongside lab experiments, the astronauts themselves have been testbeds, allowing us to make remarkable, silent advances. Eventually, when tourists take their first steps on the Moon, it will be possible thanks to what we have learned on space stations.

But my driver's question still nags: How long can we expect to wait before average people might travel to Mars? She seemed impatient to know. "Why can't I go now?" she demanded. She looked wide-eyed into the mirror again.

"Well, I think a lot of the delay has just been a lack of political will," I explained. "Those space projects like the Apollo Moon missions were all propelled along by the government. You see, America and the Soviet Union essentially thought of space as a place to display the prowess of their societies. The

Moon was the United States' response to the growing confidence of the Soviet Union in Earth's orbit. They had already launched the first dog, the first man, the first woman, and the first crew, and putting a human on the Moon was the next obvious step. If the Americans could do it first, that would make all Soviet accomplishments not so much irrelevant but secondary to our ancient dream of touching the Moon."

I carried on, explaining to my driver that, in this competitive atmosphere, there was no place for the average citizen other than to watch and cheer on enthusiastically. The people chosen to sit in the capsule had to be logical, fearless, creative, emotionally stable, physically and intellectually gifted, calm under enormous pressure, and interested only in successfully carrying out the mission. This was no standard-issue assignment, nor did politicians and government administrators dream that space missions would segue into a tourist industry.

By the time the Americans had planted their flag on the Moon, this view of space as an environment strictly for professionals was firmly in place. The US space shuttle came and went as the world's first reusable spaceship, with nary a tourist aboard. (The Soviets had their own shuttle, the *Buran,* or *Snowstorm,* which flew only once.) There were some "normal" people who did join space shuttle missions—the schoolteacher Christa McAuliffe, for example, one of those lost tragically in the *Challenger* disaster. But she was selected for the role from more than 10,000 applicants and trained extensively. The International Space Station, a product of global cooperation beginning in the 1980s, launched in 1998 with government-backed astronauts as its operators. That remains the case today.

But here comes the uplifting part. "All that really changed in 2001," I explained, "because that was the first glimmer of hope for the rest of us." What I had in mind was Dennis Tito's eight-day trip to the Russian Mir space station. A space scientist turned investment banker, Tito had earned a degree in astronautics and aeronautics at New York University and then worked at NASA's Jet Propulsion Laboratory in California. Ultimately he put his knowledge of mathematics to use analyzing market risks, ending up with a billion-dollar personal fortune. With all that cash, he formed a relationship with MirCorps, a company that wanted to take private enterprise into space by flying tourists to the Russian station. NASA was not terribly impressed by Tito's plans, it must be said. Dan Goldin, the NASA administrator at the time, thought tourist trips to space were inappropriate. But in April 2001, with help from another company, Space Adventures, Tito did finally fly into space.

Tito's mission didn't open the floodgates. With the $20 million price tag quoted for his trip, space tourism was not ready for takeoff, so to speak. And Tito wasn't totally out of place on a spacecraft, given his background in aircraft engineering. Nevertheless, this was a turning point. The psychology of space travel had changed. Tito's trip didn't devalue astronauts; the elite NASA corps was still elite. But if a sixty-year-old could spend more than a week orbiting Earth, then perhaps space was not only for fighter-pilot heroes after all. Tito showed that space tourism was possible, if not widely accessible. A normal person without years of training could fly into space, spend several days, help with experiments, and return home safely.

From a practical and organizational standpoint, of course, this was not space tourism of the kind we might imagine becoming routine. This was very much a boutique enterprise, too small to operate at scale and still dependent on government programs. Mir, Tito's original target, was operated by the Soviet and later Russian governments, as was the Soyuz that he traveled in. The mission docked at the International Space Station, another state-run facility. And Tito did some training with NASA.

"It's all government, that's the problem isn't it?" my driver asked. "The government won't sell me a ticket," she said, gesticulating toward the sky with the fingers of her right hand.

She had come to the same conclusion as some very wealthy entrepreneurs who are now trying to change this situation: what the world needed was a genuine private taxi into space. I told my driver about Elon Musk, the tech billionaire who founded SpaceX the year after Tito's mission. By 2008, the company had launched the first private rocket into orbit. NASA was so impressed that it paid the company to send cargo to space, and now more than twenty private resupply missions have gone to the International Space Station using SpaceX's Dragon class of spaceship. Another, human-rated, version of the spacecraft takes astronauts to the space station.

And in 2021, I told my driver, "the passengers on the Dragon changed from good old-fashioned government astronauts to genuine private tourists like you." She physically jumped in her seat. That mission, known as Inspiration4, featured the first bona fide civilian astronaut crew to orbit the

Earth. "In some ways, they are opening the gates to people like you," I said. "As more private companies get into space, so the technology becomes more reliable and safer, making it easier to justify sending paying tourists."

"So I'm getting quite close to it do you think?" she asked.

"Yes," I replied. "I think we all are."

SpaceX is at the vanguard of private-sector space activity, but it is not alone. The industry is growing rapidly. Jeff Bezos, of Amazon fame, has used his wealth to set up Blue Origin, a company with aspirations to launch rockets into orbit and farther afield to the Moon. In the meantime, it has achieved spectacular success with its New Shepard, a class of spacecraft that can take tourists on a short suborbital flight. The craft leaves the Earth's atmosphere, where passengers experience a few minutes of weightlessness and an unparalleled view of Earth from the blackness of space, before the capsule descends and plops them down gently in the desert. Tickets for this flight may cost more than $100,000, but this is a far cry from the $20 million that Tito paid.

Entertainment and airline mogul Richard Branson has also thrown his hat into the ring. His Virgin Galactic company has built and flown several spaceships. Virgin Galactic's SpaceShipOne, the first private spaceship, entered space in 2004. It was designed by visionary engineer Burt Rutan. This was followed by the upgraded SpaceShipTwo, but a terrible setback came in 2014, when the first version of the vessel disintegrated in flight. The accident, caused by the premature deployment of the ship's feather mechanism, which is used to decelerate before landing, took the life of copilot Michael Alsbury. The second version of the ship, however, was successful. Branson himself joined the mission, a one-hour jaunt to the edge of space. Not quite the Moon, but part of an invigorated push outward.

I thought I'd try my driver out. "So for just one hundred thousand dollars you could experience space for a few minutes," I stated matter-of-factly, to see how she would react. I was not expecting the answer I got.

"You are joking aren't you?" she said. "I don't want a few minutes. I want to go to Mars." It was as if I had denigrated her life's vision. I was pleasantly surprised.

These days, private companies have plans to go well beyond Earth orbit. SpaceX already does so and has prototype rockets to take people and materials to Mars. Other companies plan both robotic and crewed missions to the Moon. It's not easy to predict which companies will succeed; like all private businesses, many come and go. The interest of investors ebbs and flows. But we need not be concerned with which brands win out. What is important is the increasing accessibility of space. The cost of the technology required for space travel has dropped enormously, so that the opportunity is within reach of private individuals. As more firms vie for the space frontier, so they test innovative engines, fly new capsules, and put novel materials through their paces. These businesses are contributing to humankind's knowledge of how to fly into space. For every generation the dangers still lurk, but we are getting closer to the day when taxi drivers will be able to afford safe travel into space. Nearing too is the day when even Mars will be an option for the adventurous tourist, although that will be delayed considerably longer.

Indeed, while many firms focus on getting to space, there are also private enterprises thinking about where we will stay when we get there. When SpaceX was not even an idea, Robert Bigelow, an eccentric and enthusiastic real estate mogul, founded Bigelow Aerospace, fulfilling a childhood dream to have his own space program. In 1999 Bigelow Aerospace began to work on making houses in space, and in 2016 the Bigelow Expandable Activity Module (BEAM) was launched and affixed to the side of the International Space Station. The big, white, inflatable pod looks like an oversized marsh-mallow; once positioned on the station, it was expanded to reveal a prototype home in which astronauts and tourists can work and play. The BEAM was the triumphant culmination of earlier tests Bigelow had launched on Russian rockets.

Space homes are essential to the future of exploration. Real estate is less sexy than a fire-breathing rocket—to most people anyway—but you will need to spend your holiday somewhere. How grateful would you be if a friend gifted you a plane ticket to a deserted island?

All of this activity amounts to nothing less than the con-struction of a space economy. Again, more will follow the early corporate players. Do you want to walk around the Moon in a bulky, albeit functional, NASA spacesuit? No, you want a fashionable spacesuit in bright colors, with a helmet that will not occlude your face in pictures. There are already businesses working to meet this desire. Many more new industries will bloom, and every tiny thing, from the color of your visor to the food you eat in space, will be fair game for the commercial market. And with expanded efforts, businesses will realize economies of scale and de-velop improved technology, bringing down prices until space travel becomes as realistic as a holiday on another continent.

However, I wasn't going to give my driver a completely rosy view of what lay ahead. "You know that it's still not an everyday thing," I explained, "there's a little danger there, and even if you get to the Moon and Mars, you'd better be into rocky landscapes."

There is no room for complacency in all this. Nothing will change the fact that the Moon and Mars are harsh and hazardous. Radiation flares from the sun can kill instantly, and unlike every holiday you've taken on Earth, your vacation will not include free oxygen. The vacuum of the Moon and the carbon dioxide–rich atmosphere of Mars will asphyxiate you instantly. These are not places where you can wander outside on a whim and enjoy a sunset. And really, what they have to offer the tourist may be limited. There is no wildlife: on the Moon there is just an expanse of gray volcanic rock to the horizon, and on Mars, the same thing but in red. What would entice anyone here? Maybe the experience of the otherworldly. Maybe also a new way to experience the familiar. If you're on the near-side of the Moon, you can see Earth hanging overhead. Its beautiful hues of green and blue may change your perspective forever, just as they enchanted the Apollo astronauts when they first witnessed this fragile marble in the endless black void.

My driver needed no special motivation to pay a visit. "Oh, I'd just go, just go to see it," she told me. "I just want to know what's it's like to be there, you know, actually on Mars!" She gesticulated again to the heavens, her eyes scanning the sky as if she was looking for the planet. Perhaps you're like her. Or maybe you're one of those tourists who has a bucket list to complete: a visit to every continent on Earth, and why not Mars as well? That's a perfectly good reason to go.

It will be a while before traveling to the Moon becomes routine, and Mars will take even longer. Actually, spending time on Mars may be less challenging than vacation on the Moon: Mars at least has an atmosphere, and in many ways is a less extreme environment. But Mars is much farther away. A trip to the Moon could be a long weekend; Mars is a journey of over a year—that's some serious leave from work.

In hindsight, fantasies of holidays on Mars ten years after the Apollo missions were naïve. There were still so many challenges to overcome. The death of Michael Alsbury in 2014 reminds us that many such challenges remain. Space is not easy. She is an unforgiving frontier, and when lives are lost, we remember that space travel, though the stuff of dreams, is not all wishes and fulfillment. You cannot carelessly send tourists and fee-paying passengers into space, as a second-rate travel company might book backpackers onto a questionable bus. A bus in ill repair may break down and leave tourists stranded, ruining their day. A spaceship that isn't up to grade will kill people.

Still, though we attend carefully to the challenges, we should take some assurance from having surmounted many of them already. In 2002, when Musk announced his plans to build a private rocket, he was met with incredulity. Designing and building a crewed spaceship, putting it on a rocket, successfully launching it into space, docking with a space station—these operations were the preserve of complex, expensive, and expansive government operations. Not even the most gung-ho entrepreneur could do it. Countless critics thought SpaceX an absurdity. But the company, and a few others like it, have shown not only that private space travel is possible, but also that the private sector can contribute great innovations—that engineers without government jobs can be determined and imaginative enough to create something new and better. New solutions have led to sleek-looking spacecraft like the Dragon capsules that appear as though they came out of the 1968 film *2001: A Space Odyssey*. These are serious spacecraft, able to carry heavy payloads and operate safely with crews onboard. Musk even launched one of his Tesla sportscars into space in 2018, a frivolous gesture, perhaps—a marketing stunt, undoubtedly—but also a demonstration of technological capacity. The car now orbits the sun, an emblem of the confidence with which private industry has thrown itself into the expanses beyond Earth.

For now, most of us remain spectators of the human extension into space, but this does not mean we should despair. If you are of the Apollo age and find yourself disappointed that humans have not yet set foot on Mars, consider how much progress is revealed in the activities of today's commercial tourism and space-hauling concerns. Disappointment should be tempered by the realization that critical developments have come to pass, making space more accessible to more people than ever before. We are on the brink of a day when that vision of Earth from the Moon no longer appears only in the memories of astronauts or the pages of a book.

Who knows whether my driver will end up on Mars, where she will have the chance to look upward and gesture back to Earth in her accustomed way. But the mere fact that we have moved from dreams of space tourism to informed speculation about near-term possibilities is something to celebrate. Even as her cab crawls the streets of Edinburgh, she is closer to Mars than anyone ever was before.

The Ascent of Olympus (2002), painted by Marilynn Flynn, imagines an expedition team finally reaching the summit of Mars's Olympus Mons, the highest peak in the solar system. Who will accomplish the real heroic first?

6

Is There Still Glory in Exploration?

A taxi ride to Warwick University to give a lecture on life in extreme environments.

✳

I've always liked Warwick very much. Its long medieval high street, quaint but made strong by the unpretentious solidity of the castle, parts of which have stood since the days of William the Conqueror.

"People don't build places like that," my driver observed as the castle appeared on the left. He didn't have to add "any longer." I stared at the ramparts and towers. Even at the height of Victorian extravagance, few in Britain tried to build anything that could match the castle's simple beauty and geometric grandeur.

"It's true," I replied. "We build quickly, all for the moment, but never to last in the same way as that fine place." It occurred to me that perhaps we'd lost the taste for such construction altogether. "Do you think we can ever build like that again?" I asked. "Or was it of its time? Is it just that we don't have the motivation?"

"I do think we've lost something of the romanticism," my driver suggested. "Some buildings are like the castle in their modern way, but we don't go for the glory any longer."

"A bit like the old days of exploration," I offered.

My driver appeared pensive, almost melancholic. He looked about sixty-five and wore a brown tweed cap and a green jumper. Now and then, he thoughtfully scanned the horizon, seemingly searching for something. Sometimes he sighed as if fed up, disappointed even, with the whole theater of life. My comment about the glory days of exploration also had him ruminating. You hear this sort of longing whenever an obscure feat is announced—say, someone just crossed the Channel in a bathtub.

"Yes, that as well, he said," after pausing to tilt his cap. "Well, we do seem to have finished all the great firsts haven't we? We've been to the highest mountains."

"I know this seems a little strange to say," I suggested, "but what about if we leave the Earth altogether? Maybe we just need to go to other planets."

Sometimes you know you've pushed too far. I'm used to conversations about space with my colleagues, so I don't filter what I say. But others are not inured. My driver laughed and gazed into his mirror with that benevolent and quizzical look that screams, "Ah, you're one of those crazy types aren't you?" His unspoken response left me that much more persuaded that our sense of glory really has gone dormant. For many of us, there is nothing to seek out beyond the confines of planet Earth. Perhaps this is just how the human mind operates. In the future, when we have settled Mars and the Moon, will explorers sit restlessly around their habitats, longing for further adventure? Will a new energy and spirit be unleashed onto new frontiers, rejuvenating that sense of the heroic currently in abeyance?

That long trail of mountaineers, packed one behind the other awaiting their turn to step on the summit of Everest, may give you the impression that we have come a long way from the dazzling days of exploration when Earth still offered plenty of firsts. When Edmund Hillary and Tenzing Norgay first stood on the highest of summits in 1953, could they ever have believed that sherpas would spend their time picking up trash at Everest Base Camp or collecting the bodies of mountaineers who, though having grown in number, still face the unyielding demands of nature? It probably would have been inconceivable to them that the accumulation of waste at Everest would become an environmental challenge.

Even in the frozen polar wastelands, adventure has been downgraded. These days adventure means being the first to cross Antarctica on a motorbike, say. Explorers engage in arguments about whether someone's claim to an unsupported crossing of the White Continent is valid because, along part of the route, they skied a road of flattened snow conveniently made available by the efforts of a scientific program. Certainly the dangers are still there. A whiteout, a lack of planning, or an unforeseen medical situation can leave you crossing the Rubicon from derring-do to fatal predicament. But given all that has come before, and all the infrastructure that others have left behind, the edge of glory has been lost. Even the most remote places have been visited many times before, leaving the pure-minded to lament the end of the Heroic Age of Antarctic Exploration.

Yet it would be wrong to believe that heroism ever was limited to the sorts of feats the great nineteenth- and twentieth-century explorers pursued. Throughout history there have been many heroic ages, as humankind's vision has soared. Perhaps to a band of humans sitting in the valleys of Africa hundreds of thousands of years ago, the first of their number to peer beyond the trough and journey to the unknown was a hero. Those early groups who crossed Asia and braved the oceans in boats to settle Polynesia were probably heroes to their kin. But those stories were told too long ago to be remembered now. We cannot hold their characters in high regard as their contemporaries did.

And much as heroism preceded the so-called Heroic Age, so we are not really at the end of the road of magnificent feats, the age of firsts. Greater challenges call us. Despite my driver's suspicion about my mental state, he was not going to escape that easily.

"I mean, if there was a great first to be done on another planet, like Mars, something as good as Armstrong on the Moon, or Hillary and Tenzing on Everest, would you do it?" I pressed.

"Mmmm, mmmm, yes, why not?" he replied. I still hadn't convinced him I was going anywhere sensible with this conversation. My journey was short that day, and I didn't have time to try to change his mind, to enthrall him with the idea of a Martian mountaineering expedition. But since I have your attention, reader, may I try this out on you?

Take a moment to do what no polar explorer or mountaineer has done: set aside Earth. Redefine your idea of the Heroic Age so that it is encompassed not by the boundaries of Earth but by the outer limits of the solar system. Instantly new frontiers come into view, frontiers as impressive as any that could have been considered by the explorers of yesteryear.

Imagine a mountain so immense that when you stand on its summit you look not into the thin blue vapor of a familiar sky, but into space. Around you blackness reveals a glittering of stars, and at the horizon, the thin veneer of the atmosphere hugs the curvature of the planet. You are imagining the top of Olympus Mons, an accumulation of lava forming what is known as a shield volcano. Rising more than 21 kilometers above the Martian surface, two and a half times the height of Everest, the summit of this behemoth is the tallest peak in the solar system. Whoever can stand on it will have achieved something extraordinary. Hillary would have been in awe.

It would be a mistake, though, to simplistically transpose past glories onto new ones. Olympus Mons is very different from Everest, and its conquest will be another kind of affair. For instance, Everest climbers routinely rely on oxygen cannisters when they reach great heights, though some have summited without them. But the atmosphere of Mars is thin and has no appreciable oxygen, so the mountaineers of Olympus will have to wear spacesuits from the beginning of their trek at the foothills all the way to the summit, every second of their climb. Their only reprieve will be a pressurized tent, pumped with enough oxygen to provide a few hours of suit-free rest.

The hikers might start their journey from a ring of tormenting sheer cliffs up to 6 kilometers high. To climb such verticals while enveloped in a spacesuit and carrying a full complement of supplies might be impossible, even as Mars's low gravity, exerting three-eighths the force of Earth's, lessens the weight of the provisions. So the climbers may instead attempt their journey from the northeast side of the mountain, where the foothills are less extreme and access to the gentle slopes of the volcano is easier.

This is one respect in which Olympus is less challenging than Everest: instead of going up and up, you trek across craggy lava fields with an almost imperceptible slope of about five degrees all the way to the apex. There are no glaciers, unpredictable avalanches, and crevasses. But the slope does go on for a numbing 300 kilometers. There will be days and days of hard slogging among fractured volcanic rocks with jagged, spacesuit-slicing edges. Perhaps the danger will sharpen minds otherwise tormented by boredom.

The prize at the top is a giant caldera, a 60-by-90-kilometer ellipse of what were once lava lakes—the remnants of vents that gurgled liquid rock. From the edge of the rim, the explorers will have their machine-oxygenated breath taken away by a view across the mighty Valles Marineris, a canyon system thousands of kilometers long and several kilometers deep, so large that the Grand Canyon would vanish within it. From the top of Olympus, the hazy salmon sky of Mars will appear below, and here and there will be faint wisps of Martian clouds.

When they reach the top, the adventurers will have restored the Heroic Age, within a wider boundary. But one hopes they will not just gather a souvenir rock or two and then turn back, like every mountaineer at Everest. There is a lot of work to be done at the capacious caldera of Olympus, which can tell us about the history of Mars when the planet was more active, when heat and water may have made these sunken, collapsed bowls a habitable environment for life. The first people to climb Olympus would be wise to collect samples of lava and the mineral remains of the water systems that circulated from deep in the volcano to its surface. Within these samples lie clues to the ancient history of our sister planet and the reasons it descended into a long deep freeze, while our own world effloresced with oceans and life.

Mars, although a new world to its first human explorers, will present some earthlike features. Like Earth, sitting at its antipodes are polar caps. On Mars, the poles are solid water ice, covered in a seasonal layer of carbon dioxide snow. Stand close to these caps or fly above them and you will see their raspberry-ripple appearance, long serpentine reds and oranges running lengthwise through the ice. These are layers of ancient dust, blown by storms until they were trapped in snowfall. Now they call us to a treasure trove encoding millions of years of Martian history. The ripples constitute a geological time capsule that will tell us about climate variations on the planet, which themselves provide a window into the recent history of our whole solar system.

It is quite impossible to look at satellite images of these polar caps without being entranced by the idea of traversing them. Yes, you could just land a spacecraft at one or the other Martian pole, drive a core into the ice, collect a sample, and return home. But that would hardly be the point. And here is a sure difference from the past ages of exploration on Earth: we can see the Martian poles from the comfort of earthbound armchairs before humans have even been to Mars. Exquisitely detailed images, glimpsed by Mars-orbiting satellites, are just an online search away. Roald Amundsen, Captain Robert Falcon Scott, and Sir Ernest Shackleton, by contrast, had to use their imaginations. No one had ever seen the places they visited. They and the wider public could only dream of the splendid, awful isolation of Earth's polar wastelands.

Yet, if we have a sense of what awaits us on Mars, this does not mean that our polar explorers will have an easy go of it. Our intrepid team might begin their first unsupported trans-polar assault on the Martian North Pole from its edge near the Chasma Boreale, a wide valley cutting into the polar ice. From there they face a journey of over 1,000 kilometers to cross the ice. Like their mountaineering friends on Olympus, they will make their long march in spacesuits, removable only in the pressurized tent they will haul along. Good thing, too. It is tough to sleep in a helmet and body armor.

As the sun hugs the white horizon each morning of their crossing, they will not feel the crunch of fresh snow underfoot. Here the temperature is below –100°C, and the snow is as hard as concrete. The journey cannot be accelerated with the sort of skis or sled one might use on Earth; deploy either, and it will soon be torn to shreds. Instead the travelers will have to walk in warmed boots, dragging behind them a container on wheels, treads, or, better still, heated runners. As the atmosphere is so thin on Mars, the ice, once warmed, does not turn into a slushy mess but instead instantly vaporizes; the heat from the runners might create a thin cushion of air, allowing the hikers to pull their supplies relatively effortlessly.

They will trudge on for about eighty days. While they walk, they will consume food and water through their spacesuits. Perhaps the food will be in liquid form, connected by a tube into a tub of tasty nutrient broth in their cargo container.

They might have to bring all their oxygen, or they might carry a system that could make oxygen from the Martian atmosphere. Certainly it wouldn't be sensible to drag all their water, given that they will be surrounded by water ice. They could use heated rods to cut into the ice and remove chunks, warm and pressurize the ice, then filter and clean the resulting liquid to remove dust and salts, providing them with a fresh drink.

There is very little variation in the terrain of the Martian poles. Mostly it is whiteness to the horizon, with occasional patches of red and orange dust and potholes where the ice has irregularly vaporized away under the Martian sun. But with some deft ground navigation and maybe the aid of a satellite, the adventurers will find the geographical pole. There will be no whiteout blizzards to hold them fast, pinned to the ground like Scott and his team in their final hours. There will be the faint whistle of the Martian wind against the explorers' visors, but that will be all that accompanies their faces grinning with pride into the Martian desolation.

On this day, millions of miles from Earth, an event of no significance to the universe but of abiding importance to humans will occur. Here, a heroic first will be achieved. Such a journey will be symbolic, and that is the point. By the time humans can trek a Martian polar cap, we will also be able to land a rocket precisely at the pole. Indeed, it may be that our first hikers will find evidence of earlier rocket-powered journeys—a weather station or a supply drum, long since frozen solid into the surface with a bank of snow piled against its side. But never mind. This is a human story, an endeavor in the face of challenge. No matter what detractors may say, generations will be inspired by their tale of exploration and the new chapter it opens. When these unnecessary adventurers finish the second half of their traverse and clamber into a rocket for their ride home, maybe they also will haul a wealth of samples—drill cores, dust, water—that will bring new insights into the evolution of Mars, its climate, and its potential for harboring life.

In keeping with our terrestrial history, a polar crossing may be a prelude to something more: circumnavigation. Circumnavigation is the explorer's gold medal; long before the Heroic Age, explorers had in many ways already achieved bigger things by going all the way around the Earth. The ship *Victoria,* under the command of Portuguese explorer Ferdinand Magellan and his Spanish colleague Juan Sebastián Elcano, sailed across the Atlantic, Pacific, and Indian oceans between 1519 and 1522. It was not until 1979 that Earth was circumnavigated across the poles, by British explorer Ranulph Fiennes and his team. Their Transglobe Expedition began in England, headed south over Antarctica, then went north across the North Pole, and then south again, returning to England having rounded the planet.

On Mars, one could recapitulate Magellan and Elcano by circumnavigating the equator, a 21,000-kilometer expedition across nothing but desert. (That is the direct distance; in reality, the many irregularities of the terrain would make for a much longer journey.) This would be an extreme-duration expedition, requiring days to cross individual craters, dunes, rocks, and mounds. Yet the sheer monotony and danger to people and vehicles would make it a triumph of exploration.

The stories to be told after such an immense undertaking would thrill humanity.

And what about a Martian version of Fiennes's polar feat? I have long been fascinated by such a prospect. As I imagine it, a trans-Mars expedition begins by crossing the ice of the north pole, before arriving at the circumpolar dunes. From there, the expedition advances across deserts and craters, arriving at the edge of the south polar ices, where the team pauses to cherish the moment before beginning their second polar crossing. In the second half, as the explorers return to their starting point, they summit Olympus Mons. By the time they have returned triumphant to their starting point, they have covered 19,000 kilometers of deserts, over 1,400 kilometers of ice, and 700 kilometers over the highest mountain in the solar system. Now if you want a first, this is it. It would not be the longest circumnavigation—Earth's circumference is about twice as large—but the challenges make this expedition second to none. The trans-Mars team will have to endure constant enclosure in pressurized spacesuits or habitats, extreme temperatures, and vast vistas of eroded and machine-destroying rock and dust. I challenge you.

Plenty of other possibilities await the outward-looking adventurer. We might circumnavigate the Moon, or climb the ice cliffs of Miranda, a moon of Uranus. One day, someone might even circumnavigate the methane and nitrogen snows of Pluto.

For every generation of humans, it seems that the past is a hard act to follow. It would have been easy to ponder the feats of Magellan and been left in a state of helpless paralysis at the scale of what he and his sailors achieved. Yet in the twentieth century, a small band of explorers did consider themselves up to the task of equaling their forebears, so they set themselves the goal of circumnavigating Earth over both poles. The trick for each generation is to reset the limits, to redefine humanity's capacities and possibilities. Magellan could never have contemplated a circumnavigation of the poles, because these regions of the world were unknown. The expedition was not even a possibility. However, with new knowledge and technology, a challenge equal to his own opened to those who could imagine it.

Today children are born into a world where space exploration is becoming more available thanks to better tools and technologies. Rather than wistfully looking back to the age of Scott, Amundsen, and Hillary, the boundaries must be established again. We now have images of Olympus Mons and can plan expeditions across the Martian polar ice caps. We can even describe in detail an expedition to circumnavigate Mars over both poles. We can't do these expeditions for real quite yet, but maybe in a few decades. That is nothing in the long march of time. There is a grand and heroic age of exploration waiting to unfold, offering challenges as great, and greater, than anything we might try on Earth.

Centuries from now, we will revere the explorers of Earth, their stories and their courage. But our history books will also regale readers with the exploits and near-death dangers of those who ventured across some of the most forbidding landscapes in the solar system. Niles Brandrew, who summited Olympus Mons; Emily Hawkins, who first crossed the Martian

North Pole over land; Wu Wieran and the team that first circumnavigated Mars. Who are these people? What are their real names? Someday humanity will know the explorers who took us to new heights of daring and kept alive the spirit of exploration that has fired so many hearts since the first expedition that left that valley in Africa.

Mars in all its arid splendor, the great Valles Marineris canyon system cutting
across its surface.

7

Is Mars Our Planet B?

*A taxi ride from San Francisco airport to
Mountain View, California.*

✳

I had flown from Orlando, Florida, where I had been at the nearby
Kennedy Space Center attending the launch of an experiment to the International Space Station. Now I was in California to run our ground experiment—the exact replica of
the experiment in space, so that we could compare results in
low gravity and in Earth's gravity. This had been ten years in
the making, and we were excited to see the results.

The purpose of the experiment was to test the effective-
ness of "biomining" metals using microbes. Here on Earth,
microbes have been breaking down rocks for billions of years,
so they are pretty good at the job. Scientists have also tested
the process in controlled conditions, using microbes to extract copper and gold from rock. It's environmentally much
safer than pouring nasty chemicals like cyanides into rocks,
another way to get the useful elements out. What our research
team wanted to know was whether we could make the same
process work under different gravitational conditions, in the
hope that at some point we could biomine rare earth elements
and other valuable minerals from space rocks or asteroids. So

we tested the process under conditions of weightlessness and while using a spinning device to simulate Martian gravity. A couple of months later, we would find out that our little experiment had worked; it was the first demonstration of biomining in low gravity.

In the meantime, though, I just needed to go to my hotel. When I got into the cab, the news was on, another bulletin on the state of the world. My driver was silent at first, but her attention shifted from the radio as we sped off down highway 101 toward Mountain View.

"So many problems in the world. Don't you think?" she asked. She was a bubbly thirty-something, waving her arms up and down and speaking in that soft Northern Californian accent. She wore a bright orange and red T-shirt, and her long, wild auburn hair hung on her shoulders. She was one of those people whose eye color you register instantly; large eyes, insistent, staring at you and expecting answers, an almost childlike demand for attention. They were dark brown.

I agreed with her, it seemed like there was a concatenation of problems around the world, and in more downbeat moments one might be forgiven for believing that things were in a downward spiral. "It's true, a lot of disagreements, tensions over everything from oil to nuclear weapons," I replied. "Having said that, there is a lot of good in the world," I offered with a vague sense of hope.

"Well we gotta get it sorted, right? There's nowhere else to go," she suggested.

To me, this sort of comment is like candy to a kid. "You mean, no other planets to go to?" I inquired.

"Yeah, this is the best place we have. All this talk about going to the Moon to escape the Earth. We gotta solve our problems here," she said.

She had hit upon a common criticism of space exploration—not just that there are better things to do right here, but also that those who want to go into space are motivated to escape Earth because we are trashing it. As our environment is degraded and the human population increases, the obvious solution is to just leave Earth. Go somewhere else, find a new home. Set off for planet B.

No matter how often I hear this perspective, I'm slightly at a loss to understand where it comes from. Maybe some television programs, books, and other media have, intentionally or otherwise, convinced people that we should settle space because we are making a mess of Earth. Or perhaps this misapprehension can be credited to no one in particular but instead is implied by the enthusiasm of would-be space explorers. After all, when we go on passionately about establishing settlements on the Moon, Mars, and beyond, people will make assumptions. No matter its origins, the we-have-nowhere-else-to-go complaint is a most egregious one, and I want to explain why.

Let's face it, we do have problems on Earth. When I say "we," I mean the whole of humanity, although it's easy enough to identify particular problems in certain places. We have a vast population of over seven billion people, serious environmental challenges, and many political conflicts that come in a profusion of forms. Little wonder then that some people think settling other planets would be a good backup plan, particularly if things get so bad that Earth becomes, at least for its human occupants, a write-off.

Superficially, a certain logic prevails here. The view that we need a planet B seems a cold but sensible one, and certainly it is part of the argumentative repertoire of some who look forward to space settlement. Geologic history also seems to support this argument. In particular, there are those who contend that we should look for planet B because, at some point, another calamity like the one that killed off the dinosaurs will occur.

The story of that extinction is itself a fascinating one, and the more one knows about it, the more worrisome it becomes. Sixty-six million years ago, an asteroid collided with Earth and lofted so much dust and soot into the atmosphere that the planet was plunged into darkness, a so-called impact winter. The catastrophe ended not just the 165-million-year reign of the dinosaurs but about 75 percent of all animal life, a fact less often noted. The evidence for the cosmic origins of this disaster was first dug up, literally, by Berkeley geologist Walter Alvarez and his colleagues in the 1980s. He was investigating rocks from the last years of the Cretaceous period, when the mass extinction occurred. To his surprise, he found in the rocks an unusually large amount of the rare element iridium. This element happens to be most concentrated in the deep interior of Earth and in asteroids. No plausible volcanic eruption could have delivered enough of the stuff to the surface to produce the values Alvarez was finding, so he surmised that this was evidence of an asteroid impact—a cataclysmic one.

His theory rests on other evidence, too. For instance, other researchers found tiny spherical glass beads from the same period that bear witness to vast amounts of molten rock ejected by the impact and strewn across Earth. There are also large tsunami deposits from the time period found in the landmass of what is now the United States, suggestive of giant waves propelled by the crash of a 10-kilometer-diameter object at high speed. Even the tiny fragments of rock at the geological boundary have lines in them, consistent with the massive shock pressures transmitted through the ground when that 10-kilometer-diameter object struck the Earth. The energy released in such an event is enormous: we could hope to match the ferocity of this disaster only by detonating billions of nuclear weapons all at once. That is no exaggeration. In just an instant, the entirety of the Earth's surface was changed for good.

As I noted before, these collisions can seem remote in time, even primordial in character, yet Earth is very much part of the space environment, and another such strike is not just possible but guaranteed, given sufficient time. How much time? If you count the craters on the Moon and other rocky worlds in our solar system, you get some idea of how common impacts are. The number that comes back at you is something like an extinction-causing asteroid every 100 million years. But while that may sound reassuring, a couple of nasty facts lurk beneath. For one thing, it doesn't mean that we are 34 million years from the next strike of this sort. Once every 100 million years is the average frequency, not a guide to when the next incident will be. That number would still be accurate if Earth were decimated by an asteroid impact tomorrow, followed by a few hundred million years of silence that none of us would be alive to enjoy. The second unpleasant thing to know is that an asteroid need not be of extinction-causing scale in order to cause a lot of damage. In 1908 an asteroid exploded over Tunguska in Siberia, flattening about 2,000 square kilometers of forest. If that happened over a contemporary city, it could kill millions of people. Such an impact occurs about once every thousand years—again, on average. The next Tunguska blast could come tomorrow.

The silver lining in this depressing narrative, described in chapter 4, is that we can think of ways to map and divert these objects. One option is kinetic: ramming an asteroid with an impactor. This is what NASA's DART mission is doing. Clever engineers have also dreamed up ways to nudge an asteroid away from Earth using lasers to burn off material from one side. The ejecting vapor will disturb the asteroid's trajectory and hopefully send it hurtling past our planet—as long as the offending rock is caught early enough.

With all our knowledge about the extirpation of dinosaurs and the possibility of extinction events, and given that we have the technology to map asteroids and potentially deflect them, why play Russian roulette? Why wait for doom to befall us? Well, dear reader, that is a good question. The dinosaurs would be flabbergasted by our nonchalance, and so am I. Go and ask your local space agency why they don't take the threat of asteroid impacts more seriously.

Of course, even if we try harder, we may not succeed. The finest technology may fail to detect or deflect an asteroid on a collision course. And we haven't even talked about comets. As comets tend to have orbits that take them into the far reaches of the solar system where we cannot map them, and because they travel much faster than asteroids, one of them could fly in with little warning and end our party before we know it.

This is where planet B comes in. We may not be able, or willing, to do what is necessary to ensure Earth's iron-clad safety from cosmic missiles, but we could enhance our species' long-term survival chances by establishing an independent branch of humanity on another planet, a self-sufficient colony of our brothers and sisters. No matter what happens to our blue and fragile orb, they would survive. Sure, they too could be struck by an asteroid or comet; they play the same game of roulette. But the odds for civilization as a whole would be much greater. With humans on both Earth and Mars, say, the species would be protected from destruction unless a solar system–wide catastrophe occurred.

As a "multiplanetary species," humanity would be relatively immune to other disasters as well. Our interplanetary insurance policy might save us from extinction by supervolcanoes, eruptions on a scale far larger than any volcano witnessed in human history, which would fill the atmosphere with noxious gases and suffocate life in the oceans and on land. These too are no mere whimsy. The dinosaurs get all our attention, but the end-Permian extinction was even worse: 250 million years ago, an estimated 98 percent of all animal life on Earth died all at once. The best evidence points to continental-scale volcanic eruptions as either the culprit or at least an important accessory.

And it is not as though the planet has ceased to burn from within. Yellowstone National Park—a medley of bubbling hot springs and geysers, whose water is infused with a mineral and microbial kaleidoscope of yellows, browns, pinks, and oranges—is the surface manifestation of a great magma plume beneath Earth's surface. Yellowstone's restless reservoir of liquid rock erupted about 2 million years ago, again about 1.2 million years ago, and yet again 640,000 years ago. That eruption 2 million years ago was so violent that it left a crater 80 kilometers in diameter. What would happen if such a monster awoke in the present day? It would throw out volcanic gases and particles, cooling the entire Earth. The exact effect is hard to predict. It could be a disaster on the scale of the impact winter that killed off the dinosaurs. At a minimum, such an eruption would cripple the world economy.

Now, I must add that it is not certain humanity actually would be destroyed by a repeat of the impact winter or the end-Permian extinction, even were we not to take out our interplanetary insurance policy. We may not always feel like it, but we are brighter than the dinosaurs, so it is conceivable that we would devise mechanisms to cope, preventing extinction. Sixty-six million years ago, anything that survived the initial impact and resulting shockwave, fires, floods, and so on was cornered, left to survive on luck alone. In most cases, their luck ran out. Those animals that made it through—small shrew-like mammals that could burrow, eat roots, and make a living in a withered biosphere—survived and became us. Alongside them, crocodiles and the avian dinosaurs—that is, birds—would see another day. (I think it is a much-underappreciated fact that dinosaurs as a whole actually did not go extinct at the end of the Cretaceous, only most of them. The rest are alive and well in the 18,000 bird species on Earth today. I've always thought that if we renamed chicken sandwiches dinosaur sandwiches, it would spice up our lives. But I digress.)

Unlike our lizard predecessors, we can use ingenuity to survive. If the atmosphere were poisoned by volcanic emissions or impact dust, then we would be in serious trouble, but humans have endured extreme conditions before. For instance, the Inuit of Canada's far north have tolerated the polar winter for millennia. Perhaps we could grow plants in heated greenhouses and farm enough animals in caverns to keep a small band of humanity alive. Their existence would be wretched and brutish, but at least they would still be kicking. Maybe that much-diminished society would still be larger than any outpost we could build on the Moon or Mars, so that planet B would be Earth itself, impoverished and beleaguered. Slowly, but defiantly, small families of humans could start again, recapitulating the transcontinental migrations of our forebears as they spread across the Pacific, Asia, and Europe. Perhaps the survivors or their descendants would link up and define a second great flourishing of humans, the post-impact civilization.

All this, however, constitutes a pretty big risk. Even if we had a computer simulation of an impact or volcanic disaster, we could not accurately predict whether our society could survive. The social and physical rearrangement wrought by such a catastrophe would likely send us spinning into chaotic and unforeseeable outcomes. Perhaps our species would teeter on the edge of extinction, so that a small perturbation here, an unpredictable nudge there, would decide the difference between survival or extinction. In spite of all our technology and know-how, the future of our Earth-based civilization may be decided by a roll of the dice just as was the fate of the dinosaurs.

So we return to the insurance plan. How effective would it be? At first, we would be able to settle only a small number of people elsewhere. Maybe a few tens or hundreds. But even if you went wild and dreamed up cities of millions on Mars, that's still a small number compared to the more than seven billion on Earth. And let's face it, even with a million people on Mars, over seven billion dead people on Earth would be a disappointing day.

But let us, for one moment, contemplate the mere possibility that we have the technological capacity to establish an offshoot of our civilization on another planet. We don't at the moment, but if we were truly invested in the idea, we could have that technology in just the next decade. Immunity to a planetary-scale disaster on Earth is within our grasp. Why don't we rush at this opportunity, seize it, and become the first Earth-derived multiplanetary species? I think that is a goal worthy of our capacities.

But we should be clear-eyed in its pursuit, lest we fall into the trap my driver did. In this trap, we mistake an insurance policy for an escape hatch. These are not at all the same. An insurance policy kicks in when disaster strikes; an escape is our last option when we have caused the disaster *ourselves*.

Look at it this way: no one wants to use their insurance, and not just because lawyers and claims adjusters are tight-fisted when policyholders come knocking. More importantly, no one wants to suffer the hardships they've been insured against. The same is true of our planetary insurance policy.

No matter how excited you might be about settling other planets, no one in their right mind would prefer them to our own. Every other object in our solar system is far less hospitable than Earth. Need I list all the awful things about the Moon? High radiation, no liquid water, barren gray landscapes that run to the horizon. No life, no sounds, temperatures that oscillate between freezing and boiling. If you think the comparatively clement Mars is the answer, this most earthlike of worlds in our solar system has an average temperature of –60°C, an asphyxiating atmosphere, toxic soil, extreme radiation, and no visible signs of life to temper the endless reds, oranges, and browns of the dust-filled volcanic landscape.

The simple point is this: the environment of Earth, even in its most dilapidated state, is better for humans than that of the Moon or Mars. To think of the Moon or Mars as a secure second home that allows us an escape, just in case we make a disaster of this world, is a terrible error of judgement.

As long as Earth has the natural ability to support us, the multiplanet plan should be a last-ditch one. We should seek to establish branches of humanity out in the solar system so that we can bring back home all the benefits of space, from resources to energy. And in the process, we will make ourselves less prone to cataclysm in a way the dinosaurs could not. But we should never be in any doubt that, barring catastrophe, Earth is the best planet we have for the foreseeable future.

A much darker problem lurks in the shadows for a society that sees in other planets a snowbird's second home, a beachside Florida condominium where a Michigander might hold a January barbecue. Such a vision encourages a casual contempt for our planet. Who needs it when Mars awaits? I do not think many people who support space programs genuinely hold this view. Even those who want to build a multiplanetary humanity typically see this aspiration as a back-up plan, not a deliberate scheme for fleeing an Earth used up by insouciant consumers. But as my taxi driver's comments suggest, the purpose of a planetary insurance policy is not widely understood. I do not blame her for thinking as she does. One could be forgiven for believing that space explorers are looking to move out of their first home rather than just get an insurance policy.

If you, too, hold this opinion, remember that you probably have insurance of some kind, even if you don't want to use it or pay for it and even though you are not going to set fire to your house or car or fine musical instruments or grandma's keepsakes just because insurance could make doing so less financially painful. In the same way, there is no contradiction between caring for Earth and creating an insurance policy for the human species. No matter how much we curb pollution and harden our societies against climate change and rising seas, no matter what we do to establish peace among nations, and irrespective of how we protect this planet from the astrophysical violence that is a fact of life in the universe, there remains the possibility that, in an instant, the finely balanced Earth system will be thrown off kilter and humanity will perish through no fault of our own.

Our multiplanet insurance policy might not pay out. Even our most optimistic designs might not save civilization from an end-Permian-like extinction, for it could be that the Mars colony cannot hold out long enough as Earth is slowly restored to livability. Nevertheless, why not at least attempt a multiplanet future if we have the ability to try? I think there is merit in this motive for space settlement.

So yes, Mars could be planet B. But planet B is not a seaside condo. Planet B is a hedge against the most negative prognosis for our future—extinction. Planet B isn't for sunbathing while your friends back home chop ice dams off their gutters. It's for eking out a life while Earth regenerates after catastrophe. And planet B makes no sense unless we build it while also tending to our Eden, for there truly is nowhere else like it in the solar system.

This 1899 image was produced by double exposure, but we need no clever photography to find ghosts. That is one of the lessons of quantum physics.

8

Do Ghosts Exist?

*A taxi ride from Edinburgh Airport after a scientific
trip to China.*

✳

It's not often that a conversation about the weather triggers deep
thoughts about the nature of the universe and of our existence. On
this day the conversation came with the jetlag and exhaus-
tion that accompany a long flight from Beijing. Perhaps my
weary brain just needed something to latch onto, and that's
what it did when my driver said something that reminded me
of a matter that had occupied the forefront of my mind for
some time: a truly physical appreciation of what the world is
really made from.

We were heading out of Edinburgh Airport toward the
city bypass when my taxi driver took the initiative to start
the conversation. He was probably in his fifties and wore a
thick, brown jacket with a high fur collar. Rounded glasses, a
balding head, a bolt-upright deportment, and a mild imperi-
ousness about his tone gave him the aura of a well-informed
schoolmaster.

"Funny old weather right now," he said. I didn't really have
an opinion on the matter, as I had been gone for two weeks,
giving lectures and talking biology and space exploration in

Beijing. I had been invited by colleagues at Peking University. Amid scientific seminars, I had taken the chance to give a talk at the Beijing Planetarium to an enthusiastic crowd of China's younger generation of aspiring space explorers. The December cold had been crisp but refreshing. I asked my driver to catch me up on what I'd missed.

"Well, it's not like it seems," he explained. "You can never know. See those clouds? They look all flat and gray, like it's about to snow, but then the next thing you know, they open up and it's warm and sunny. Yet yesterday it was raining. You know, you can see the weather on TV, but they never really know. And when you drive around like I do, you can never tell what's about to happen. Things aren't what they seem."

Things aren't what they seem. It's a relatively benign, uncontroversial thing to say. But within it, several thousand years of fractious and divergent thought resides. Is what we see real? When you look out of those two little spheres stuck in your head, and your brain processes the information pouring into them, do you see things as they really are? Could it be that the whole edifice of reality is an enormous illusion? Ancient philosophers had a penchant for exploring this problem. More recently, scientists, screenwriters, and imaginative people of all stripes have wondered whether we might all be living in a computer simulation programmed by extraterrestrials.

In some respects, the truths that scientists have revealed about the universe are actually much stranger than the prospect that we are characters in an alien computer game. For example, what would you think if I told you that I not only believe in ghosts, but I know that they exist? I have little doubt that you would be intrigued, and if you're a scientist, maybe appalled at my gullibility. Yet exist they do. No, I am not referring to hauntings by dead ancestors or other supernatural entities. I'm talking about everything, including you. To understand this extraordinary claim, we need to know something about how we perceive the world around us.

Plato famously compared humanity to a group of cave dwellers. All they see of the world beyond are the shadows projected onto the cave walls when something happens to pass by the mouth—mere inklings of the complexity of reality. But the methods and instruments of science release us from the cave; with the liberty to explore, Plato's troglodytes have gained some grasp of how physical reality is structured. Always a limited grasp for sure, but we are not quite so chained to ignorance as Plato suspected. What we have discovered is something far stranger than Plato could have dreamed. If one could get in a time machine and travel back to ancient Athens to reveal all, I suspect the great philosopher would be dumbstruck by how spot-on his tale was about our perception of reality, but I am also sure that he would have been astonished at how weird the underlying structure of the universe has turned out to be.

To the ancients, the world seemed reassuringly solid, as it probably does to you. When you picked up this book, your fingers clenched it with a predictable certainty. You lifted it from the shelf or table knowing that it would follow the motions of your hands until it was in front of your eyes. You opened it and, instead of seeing through it, your gaze rested on the black text written on robust sheets of paper, themselves held together in a satisfyingly rectangular chunk of matter.

The ancients had the same everyday experiences and followed them to an important conclusion about the world: that it is constructed from tiny lumps of matter. Everything from books to horses to chairs, they reasoned, was built from these lumps, stuck together. For sure, the Greeks, and everyone else for that matter, understood life as different from inanimate objects, a distinction that placed humans and the rest of living creation at a comfortable remove from mere chairs and books. But whatever that distinction was, objects like you, me, and your sofa still had the same solidness to them because all of them were constructed from the same physical stuff.

The person who most influentially described that stuff was the philosopher Democritus, who theorized that everything in the universe was constructed from indivisible particles. Democritus's ideas were so fetching that they still held sway millennia later. Even at the turn of the nineteenth century, John Dalton, a chemist who did much of his work in Manchester, England, proffered a model in which physical reality comprised tiny, solid spherical balls. Dalton called these balls atoms, from the Greek *atomos,* meaning indivisible. Each element was said to be constructed from its own type of atom, with compounds like table salt made of different types of these miniature rigid entities glued together. Although Dalton brought a more modern, chemical flavor to his theory, he followed closely in the footsteps of Democritus. Both thought they had hit upon the final irreducible particles of the universe.

A century later, the daily perception of solidity remained, even as the scientific model of the atom changed dramatically with the discovery of electrons. In 1897 J. J. Thomson, another English scientist, was experimenting with cathode-ray tubes, an invention that would eventually find itself at the heart of decades of television screens and computer monitors. By applying a current to a negatively charged electrode and studying how the particles given off were changed by magnetic fields and electrically charged plates, Thomson was able to show that the particles produced were far smaller than a whole atom and must be fragments or parts of an atom. Even more tellingly, changing the electrode materials from which the particles were encouraged to emanate made no difference to their behavior, showing that they had universal properties. Thomson had stumbled across subatomic particles common to all elements, proving both that atoms were not indivisible and that elements were not built of atoms unique to themselves. Not only that, but the fleeting character of electrons suggested that atoms were not solid—that there was something ephemeral and fluffy about their composition.

This intuition was accurate, but it could not displace the everyday experience of solidity we all know. These errant little fragments—the electrons—were therefore imagined to be firmly embedded in a cloud of positive charges that kept them in check. The atom transmogrified from a simple hard sphere into a "plum pudding"—not indivisible and uniform, but a spongy dessert. Now the atom was a loaf of positive charge with negatively charged raisins embedded throughout, so that the whole had no net charge. But *whole* is the operative word here. If the atom was no longer indivisible, it was still firm, with all its parts bound tightly together.

How could it have been otherwise? After all, the objects around us are solid, and so are we. Hold your hand up in front of you and you'll notice two of its most obvious attributes. First, it's not easy to pass objects through it. You can if you try hard enough, but you'll end up in hospital. So surely we are bona fide solid matter. Second, you can't see through your hand. If you hold a powerful flashlight on one side you might see the glow of light on the other side through the fleshy translucence, but the dim, murky light strengthens our confidence that we are made of rigid stuff.

That confidence, which had held fast for millennia, had begun to erode thanks to Thomson's work, but it would take another generation to really break down. A key agent of change was the physicist Ernest Rutherford, a student of Thomson's who sought the secrets of the atom in a piece of gold. Rutherford dangled a thin sheet of the metal in a vacuum and fired toward it a beam of alpha particles—a type of radiation that he had discovered earlier. Alongside him for the experiment was Hans Geiger, who had built a contraption that could measure the alpha particles, and Geiger's student Ernest Marsden.

These alpha particles—actually a positively charged form of helium containing two protons, two neutrons, and no electrons—streamed into the gold sheet. As they did so, Geiger and Marsden used their machine to count the number of particles that penetrated the foil. They found something remarkable: most of the particles went straight through the gold, but a tiny proportion did not. Instead, they flew off at large angles or came straight back in the direction of the emitter. The only way to explain these observations was that something had repulsed the particles. That something had to be positively charged, as identical charges repel each other. But why were just a few of the alpha particles—positively charged helium—deflected, while the vast majority of them flew straight through the gold foil? How could alpha particles rip through solid gold?

The best explanation was that the gold wasn't solid after all. Its component atoms contained a positively charged constituent, the nucleus, but the nucleus was so small compared to the size of the atom as a whole that alpha particles rarely interacted with it. When Rutherford did the calculations, he found that the nucleus was about one ten-thousandth of the size of the atom, implying that well over 99 percent of the atom's volume was empty space as far as alpha particles were concerned. In other words, the atom was no plum pudding after all—not a positively charged medium dotted with negatively charged electrons, but a positively charged central node surrounded mostly by nothing at all, with a few electrons whizzing around.

Rutherford's new model of the atom, announced in 1911, was revolutionary, but it left intact certain earlier beliefs. In particular, Rutherford had no reason to question the solidity of the nucleus or of the electrons themselves. At the same time, Niels Bohr, working in Denmark, was deepening our knowledge of the electrons in ways that also seemed to lend credence to old ideas of solidity. What Bohr found was that electrons could possess only certain, discrete amounts of energy—say, one unit of energy or ten, but not any amount in between. It would be as if you could either sprint or walk slowly, with no intermediate speed. Although this discovery might seem esoteric, it had great consequence, fleshing out the new vision of the atom. Bohr's finding also suggested that the locations of the electrons surrounding Rutherford's nucleus were not random; rather, electrons orbited the nucleus at a distance determined by the electrons' particular energy levels.

This mental image of the atom aligns perfectly with our ideas of how the planets orbit the sun, offering a beautiful resonance between physics at the largest and smallest scales. The human mind loves these sorts of alignments, which bring elegance and structure to our knowledge, in this case pointing to a consistent design throughout the natural world, from the cosmic to the atomic. Here, then, was evidence that the atom could be touched, seen, understood in much the same manner as everything else around us.

However, as scientists have come to learn, nature doesn't care about our neat stories. Bohr was right about the discrete energy levels of electrons, but other experiments would show that those tiny particles did not have such tidy orbits after all—that the atom was not a solar system in miniature. Indeed, the electrons not only lacked rigid orbits, but they also weren't even themselves rigid.

This earth-shattering discovery came courtesy of the French physicist Louis de Broglie, who showed that electrons have a split personality: sometimes they behave like the tiny balls of solid matter that fit so easily our ancient perceptions of the world, but sometimes they behave like waves rippling across the surface of a pond. Here was a great challenge to ordinary views of matter, for we do not perceive the vibrations of waves and the discrete existence of particles as the same thing.

Still, one could convince oneself that an object might behave like a particle sometimes and like a wave at other times. After all, water is fluid at high enough temperatures and rigid at low enough temperatures. But stranger things were afoot, as scientists soon discovered that electrons are not sometimes particles and sometimes waves. In fact, they are both at once, and you can bring to the fore either of these attributes depending on the sorts of experiments you conduct.

This was the birth of the quantum vision of the atom, whose intricacies were pried open by the German physicists Werner Heisenberg and Erwin Schrödinger. One of the implications of the quantum theory—truly harrowing to partisans of the old solidity—was that you couldn't say where an electron was at any given time. If you prodded and poked it, it would appear to stop dead in your apparatus, apparently looking like it was in one particular place, like your chair or your table. But this turned out to be an artifact of the experiments. In reality, the entities were spread out in all sorts of places around the nucleus, and all we could say was that an electron had a probability of being in one or another spot at any one time. It would be as if you asked me where I was located, and I replied by telling you that there is a 50 percent chance that I am at Edinburgh Airport but also a 50 percent chance that I am in my office. At the scale of matter we experience, such a statement would be cause to worry about my mental health. But in the quantum world, this is perfectly normal. Electrons don't move about in well-defined orbits but rather occupy ghostly probability fields around the nucleus. They can be pinned down when you go looking for them, but otherwise are nowhere in particular—just in all sorts of places with a certain chance of being in any one place.

The implications of this view are easy to ignore—nearly all of us do just that, nearly all of the time. But the moment you ponder it, you realize that the quantum theory demands an astonishing revision to our understanding of what matter is. Consider the following observation: that person you see standing at the bus shelter or walking out of the grocery store is made up of atoms, the volume of which consists almost entirely of ghostly electron probability fields. Of course, don't confuse this strange aspect of the quantum scale with what's going on at the scale of everyday existence, as some pseudo-scientists love to do. Your friend at the café really is there sitting in front of you, not dispersed in different places. Still, the electrons around every one of the atomic nuclei comprising her cannot be located with precision. Her material self is mainly a fuzz of electron probability fields. In other words, most of her volume is an apparition. She is truly a ghost, and so are you.

Why is this at odds with our everyday experience? Let's return to those two features of your body that seem so familiar. First, it's not easy to pass things through your hand. Despite the fact that you are a smudge of probability, there are enormously strong repulsive forces between atoms as they get very close together. Like-charged electrons in different atoms repel each other, and so too the positively charged protons, such that objects do not simply pass through each other, giving them the appearance of a stubborn solidity. Even if we push hard enough, earning ourselves a trip to the hospital, we still have not conquered solidity but merely separated one solid thing into multiple.

The illusion of bulk is given further force by the other characteristic we identified—the apparent opaqueness of solid things, and many liquids besides. View your hand under the light and what you are observing is trillions of tiny particles, photons, streaming off your hand. They might have begun their journey from your lamp, but when they strike the atoms in your hand they are reflected, eventually (actually, very quickly) entering your eyeballs and from there your optic nerve. Now this is itself a quantum rabbit hole, because photons don't just bounce off atoms like a billiard ball on a snooker table. Rather, photons are absorbed by the atom's electrons and thrown out again. The details of how light is reflected at that microscopic scale need only concern a quantum physicist; from our point of view, what matters is that this habit of atoms, the habit of spitting photons back out again, gives matter the appearance of being solid.

It is those repulsive forces of the ghostly material and their reaction to the stuff of light that has so beguiled us, convincing us of the essential continuity of material things. So potent is this mirage that we can't actually experience the world any other way. Still, you can teach yourself to think differently. Peer behind the façade of your friend at the café. Pierce the veil of the photons and repulsive atomic forces and view her as a ghostly apparition of trillions of tiny invisible nuclei surrounded by ethereal probability fields. I promise that once you have done this three or four times with all the force of your imagination, you will never see the world in the same way again. Not even an apple will seem quite as it once did.

I can't do better than to draw your attention to the lecture Edward Purcell delivered when he won the physics Nobel Prize in 1952 for the discovery of nuclear magnetic resonance, a phenomenon now used to probe the molecular structure of bacteria and the interior of human bodies for purposes of medical diagnostics, among other uses. "I have not yet lost a feeling of wonder, and of delight, that this delicate motion should reside in all the ordinary things around us, revealing itself only to him who looks for it," Purcell told the Nobel assembly, referring to the resonance he had found. "I remember, in the winter of our first experiments, just seven years ago, looking on snow with new eyes. There the snow lay around my doorstep—great heaps of protons quietly precessing in the Earth's magnetic field. To see the world for a moment as something rich and strange is the private reward of many a discovery."

Of course, to view the world like this does not necessitate that one make a major discovery. Sure, it's something wonderful to be the *first* person to see what lies around you in a completely new way. But any of us can look at a pile of snow and for a moment see tiny atomic particles spinning and gyrating in a cold winter morning. Purcell may have taught us to do so, but he is not the only one with this power of sight. That power is the fruit of scientific labors. With its nourishment, we can climb outside Plato's cave and see things as they really are. What we find out about the world does not always shatter our perceptions, though it might. At the very least, it explains what we experience. In the aha-moment, we realize not that we have been living a lie, but rather that what seems simple is in fact a spectacle, a theatrical display that distills an intricate underlying reality. We spend every day in this theater.

That's why I think Plato's cave analogy is remarkable even 2,500 years later. He saw that the shadows cast on the cave wall are not a deception or even a delusion. They are a real phenomenon, but behind them lies a series of other realities: the people walking past, the light they interrupt when they do, the remaining light that passes to the cave wall. The shadow we can perceive is a result of imperceptibilities that are no less a part of our world in spite of their inaccessibility. In some sense, we have left the cave and seen the people passing, but now we understand that they themselves are an optical illusion, another type of shadow: the reflection of the photons, the repulsion of atoms, all of which cast a different image in our minds. Yet with the tools and methods of science we have been able to dissect this new distortion and see our ghostlike form for what it is. But there should be no complacency; we should continue to wonder what realities about the universe might lie behind even this view of ourselves.

There is another perspective I'd like you to contemplate. One might think of this as the next tier of strange thoughts, but if you have some time to indulge, I'd beckon you to try this out. You might well have previously considered how remarkable it is that an intelligence exists on this planet. We contemplate our surroundings, think about the origin of the universe, and wonder about life elsewhere in it. If you dwell on that thought for a while, you realize that it is a heady one, replete with potential. Now fuse this thought with the ones I encouraged you to ponder on before.

Consider for a moment clouds of matter that are more than 99 percent nothing, just wisps of electron probability fields dispersed on a planet that itself is nothing more than an agglomeration of protons and neutrons speckled throughout a vast sea of electron probability clouds. These ghostly electron clouds communicate with one another and wonder whether out in the vacuum of space there might be other electron probability fields that interact and communicate. The ghostly clouds use the energies exchanged between the probability fields to compute, visualize, and predict the nature of the universe they find themselves in. How remarkable it is that such a thing is possible—that creatures that are not creatures at all, but rather probability fields, can know anything. These clouds of probabilities assemble other electron probabilities that take the form of particle accelerators, which are used to crash together and study subatomic particles, and giant radio telescopes that collect photons from the distant regions of the universe. The living world, the whole universe, is merely the machinations and interactions of particles and their probabilities.

When I first confronted the ethereal nature of reality in a very conscious way—and I mean in a way so concrete that it was often on my mind when I was doing everyday tasks—I was struck with something permanently wonderous. The feeling never left. I still enjoy walking along the street imagining my fellow pedestrians as they really are: ghosts going about their business, my fellow probability clouds, mostly nothing. Am I no longer sane if I find a cloud of electrons pretty or if I detect the quantum probability functions in a smile? Wouldn't it be fun to annoy a cloud of electrons, just for the entertainment of seeing a vacuous collection of probability functions getting angry? I try to desist from such behavior, though, because I can also revel in the absurdity of etiquette between collections of probabilities. The idea of subatomic entities being nice to each other has its own enjoyable ridiculousness. Overall, I've concluded that it's worth allowing your own set of empty space and probability functions to feel something for others because, otherwise, reality may be difficult to bear.

It is easy to get distracted by the excitement of looking for alien life out there, and surely contact with any entities beyond our home world would be scientifically momentous. But we should never underestimate what we can find out about life and the universe by peering into ourselves. In revealing our own ghostly form, physics has shown us that we are stranger than the most alien creatures ever imagined by science fiction writers. When we look inward, we find aliens within ourselves.

What motivates an alien? If we haven't yet encountered intelligent extraterrestrials, perhaps it is because they wish to observe rather than interfere with
us—like tourists visiting a safari park.

9

Are We Exhibits in an Alien Zoo?

A taxi ride from Swindon railway station to the UK
Space Agency at Polaris House.

✳

I didn't know Swindon well. It was the home of the United Kingdom's science research councils, but I had never really been familiar with the place. As I jumped into the back of the taxi and we sped off through the roundabouts and under a bridge, I thought I'd ask my driver.

"Swindon. What do you think of Swindon?" I asked.

My driver giggled and shifted in her seat.

"I like it here," she replied. She ruffled her green leather jacket as if to signal her pride in the place, sitting up in her seat and adjusting her permed black hair. She was well-dressed and had something of the 1980s about her. I guessed she had spent her teenage years in that era and in a place rather like Swindon.

I had no reason to disagree with her view. It was a gray, unprepossessing day, but the town was pleasant enough. There were a few people milling around outside a pub, and a crowd gathered at the edges of a large food market. A teenage girl stood next to the tent door, munching on a sausage roll while her friend braided her hair.

I was on my way to chair a committee reviewing grant proposals, and my mind was fixed on the many documents that I was about to read. As a scientist, this is one of those community tasks one should do because others make the effort to give their time to review your own pleading attempts to get your research adequately supported. But it's not the most energizing task, so my mood was not one of levity.

"Anything exciting for your day?" my driver inquired, pulling me out of my trance.

"Reviewing grants for the UK Space Agency," I replied. "I wouldn't describe this as exciting, but it needs to be done and, truth be known, it's actually quite interesting to read about all the stuff people are doing. You know, things like looking for life on Mars, building instruments to study the Martian atmosphere. It's a review panel for space exploration."

"Sounds quite exciting to me," she rebuffed. "You can't scoff at a job like that." It occurred to me that she wasn't wrong. "But Martians. I hope you don't find them," she added.

"Why?" I wondered. "Don't you think it would be awesome to find some life on Mars?"

"I've seen *War of the Worlds*," she continued, "and we all know what happens to us. Sometimes you shouldn't wish for too much. They could be dangerous."

Aliens might indeed be malcontents, and certainly popular culture presents them this way. Watch almost any film about aliens, and they probably turn up to Earth in impressive ships and with questionable intentions. On screens in 1996, US fighter jets were scrambled to discourage an alien race from blowing up the White House, resulting in an unconventional Independence Day. In 1979 Ridley Scott gave us *Alien*, about a superefficient predator that spawns its young in a hapless victim's stomach. "Get away from her, you bitch!" Sigourney Weaver cries at the alien attacking one of her crew. It's not the sort of language I would use in alien company, but there you go. I was not surprised, then, to meet a taxi driver who regarded alien contact as potentially dangerous.

In the halls of academic institutions, too, the risks of contact have not been taken lightly. Serious scientists ask whether it is sensible to transmit radio messages into space announcing our presence or encouraging aliens to visit us. What happens if "Hello, we are here on Earth" is lost in translation as, "Our planet is suitable for a complex intelligence like ourselves; perhaps Earth would be a great place for you to colonize!" Should there be an international protocol, some agreed-upon procedure, for transmitting messages to aliens?

It might seem that such concerns are inflated. After all, what chance is there that aliens exist? And even if they are out there, do we really believe that an errant message would doom us? Not only that, but we have been spilling our radio messages into space since the 1920s, so perhaps it is too late to do anything now. These were not deliberate attempts to communicate with aliens, but our transmissions may still reach unintended ears. As these transmissions dissipate into space, they are reduced to a crackle in accordance with the inverse square law, which says that every time the transmission distance doubles, the intensity of the signal is reduced not by half but by four times. Nevertheless, if extraterrestrials possessed a sufficiently powerful receiver, they could listen to our first radio broadcasts from as much as 100 lightyears away. Creatures on any planet orbiting Zeta^2 Cancri, about

83 lightyears away, could right now be hearing the rantings of Adolf Hitler at the 1936 Berlin Olympic Games. I hope they are suitably unimpressed.

Before we get too anxious about the prospects of inadvertently irritating some malevolent race of aliens, perhaps we should find ourselves some aliens to be concerned about. "I see your concern about them being dangerous," I said, "but do you think there are any out there in the first place?"

"Oh, I think there are, yes, for sure. There must be, right? There are so many stars. There must be. We'd be crazy to think we were the only ones," she replied.

That same thought occurred to Enrico Fermi, one of the twentieth century's great physicists and the inventor of the first nuclear fission reactor. Fermi gained a reputation for cooking up short, pithy questions that had no easy answers but would tantalize the intellect. Among his best known: Where are all the aliens? If you think about it, it is bizarre that we haven't come across them. In just the last hundred years or so, our civilization left behind horses and carts and transformed into a spacefaring people that has walked on the Moon. If we could do that in a hundred years, what could an alien species do with a million? Fermi took for granted that, if there are other civilizations in the galaxy, then surely some will be older than ours and therefore more technologically sophisticated. Given enough time *some* alien must be able to achieve interstellar flight. So why isn't contact with aliens a common phenomenon? Why don't aliens regularly land in Edinburgh, chat with the locals, maybe taste some haggis, and drink a nice cold can of Irn-Bru?

This thought-provoking question became known as Fermi's paradox, though that is a misnomer. There is no logical contradiction here; it may just be the case that there are no aliens out there with which to communicate. It might have been better to call this question Fermi's mystery. Whatever we call it, though, Fermi's paradox must be addressed before we start worrying about aliens with malicious plans.

My driver was getting at one dark response to Fermi's conundrum. Imagine that the universe does contain a vicious creature on the loose, perhaps something akin to Ridley Scott's alien. It travels the galaxy seeking others to consume, destroy, or subjugate. Like an over-flamboyant boar stalked by a wolf, the civilization that shouts loudest will draw scrutiny and is more likely to make the list for an unpleasant visitation. The lesson here would be twofold. First, silence is survival, so maybe we're lucky we don't have greater capacity to send out loud noises. Second, if silence is selected in an evolutionary sense, sparing civilizations the ravages of predatory aliens, then we are unlikely to hear from our fellows out there in the galaxy who did decide to make themselves heard. That might explain why aliens have never landed in Times Square: species that remain quiet keep out of harm's way, while those that go gallivanting around the universe are attacked, their efforts to make contact proving to be their undoing.

But, really, how seriously should we take this idea? On the one hand, it is hardly implausible that aliens are a marauding sort. After all, we humans are certainly aggressive. We have enough atomic weapons to destroy every city on the planet;

if anyone has a right to be gobsmacked at alien hostility, it isn't us. Yet I think there are reasons to question the notion that alien beings out there have evolved to keep quiet lest they attract some awful cosmic slayer. Despite our own tendency for conflict—and despite the possible influence of Darwinian competition on this behavior—the notion of a rampaging apex alien does not seem very plausible. What would be its motive? A mindless tour of destruction through the galaxy would seem rather pointless. Even humans, as destructive as we are capable of being, would not likely launch missions into interstellar space to kill other races, unless they are a direct threat to us. Even if we did have a strong motive—say, to evict an alien species from a planet we considered a possible second home—we would think twice, given how hard it would be to remove the aliens without damaging the biosphere in which we wish to live. If we cannot completely discount the dangers of contact as the explanation for Fermi's so-called paradox, it is nonetheless difficult to imagine a realistic biological urge that would drive a species to galactic levels of maleficence.

A more plausible twist on alien aggression is that they would destroy themselves before they reached us. As we are capable of such ruinous error, why not others? A civilization technologically advanced enough to reach interstellar maturity is also a civilization capable of self-harm. Indeed, the precise technology required to get into space—rockets—also happens to be the technology used to hurl bombs across a planet. The ability to bring planetary-scale devastation to your own world is therefore indefeasibly built into the capacity to spread into the cosmos. Perhaps these are the shoals on which alien societies capable of contact have crashed, their interstellar goals forestalled by war at home.

It may be that the danger preventing our alien encounter is the danger that aliens pose to themselves—a danger that, again, humans are only too able to appreciate.

Could it be that merely thinking about Fermi's paradox causes the paradox? Here I was, having an exchange with a taxi driver about angry aliens. If you think too hard, you could worry yourself to the point where you agree that attempting to communicate with aliens is a bad idea. Sure, I cannot think of a good reason why an alien would go around exterminating other species, but you never know. Better safe than sorry. And the aliens might, understandably, feel the same way: they are at least as sophisticated as we are and therefore are just as capable of an overabundance of caution. Thus there may be elsewhere in the galaxy a group of green-tentacled octopods sitting around chatting about Zog's paradox. You know, Professor Zog, the famous biochemist who speculated about why aliens have not visited Naknar 3.

If every species is doing this, then Fermi's paradox is self-fulfilling. All these paranoid species, all ruminating about the possibly disastrous consequences of making any noise. The real disaster, then, is that none try to make contact, even those who can—that every creature out there has followed fear into solitude. Now if that's not a real paradox, it is at least a tragic irony.

What about a more entertaining possibility? "If they are out there and we don't see them," I said to my driver, "perhaps Earth is just a zoo, you know, where the aliens take their kids at the weekend to see the strange animals, maybe have an alien ice cream, and gawp at the funny noises the humans make?" She looked in her mirror to make sure that I wasn't now making fun of her. I wasn't. It's a serious question.

"I think they'd get really bored watching us," she suggested. "They'd be better off going to a real zoo." Now this was a perspective I had never considered before. The aliens observe us from afar, but they find themselves distracted and fascinated by penguins and pandas and spend the afternoon observing Edinburgh Zoo or some wild place. That would be alien irony, and fair enough: we should not presume that ours is the most interesting species on the planet.

The exchange with my driver was slightly surreal but not silly. It pointed to a plausible reason why we may never have heard from the aliens out there. For, if angry aliens can produce a silent universe, so can beneficent ones. Maybe the aliens are concerned for our well-being and cognizant of the possibility that revealing themselves might disrupt human cultures with injurious consequences for our species' development, so they keep their distance. Like a human who surveys an ant colony from a safe remove, the aliens might be watching with fascination as our biological and social evolution unfolds. They take notes, they observe from different angles, and they wonder—but they never intervene. Like a planetary-scale zoo, Earth exhibits its fauna and flora under an intergalactic regulation that prohibits feeding the animals. Only when we achieve interstellar flight and head out into the darkness are we to be admitted to this audience of zoo-goers. Who knows, maybe our zookeepers have been doing us a favor all these years, keeping the predatory aliens at bay so that the peaceful ones can watch Earth and learn whatever interests them.

These conjectures about the character and motivations of aliens can help us think through Fermi's enigma. But at the end of the day, it may be the case that there are no aliens out there with whom we must contend, whatever their proclivities. Or perhaps the technological challenges of crossing vast interstellar distances—whether with a spacefaring vessel or an interpretable communications transmission—are simply great enough that the chasm between ourselves and even sophisticated aliens can never be crossed. These more subdued thoughts occurred to me as we turned into a roundabout surrounded by a rotunda of leaves and unkempt branches. I put the possibility to my driver. "It's possible that even if they wanted to get here they are so far away that they couldn't," I suggested. "It's just too difficult."

"Good. I'm safe then," she sniggered. Most of us who think about alien life are rather disappointed by the prospect of never encountering others out there. But here was a reasonable person relieved to be rid of an interstellar problem. It may be that we alien enthusiasts will have to learn to live with our disappointment.

There is something to be gained from such sobering possibilities: humility. In modern times, we have lost the taste for humility. Our civilization has been gripped by a sense that all problems are solvable. With the germination of the scientific method in the seventeenth century came a new sense that we could answer the hardest questions. That confidence was only intensified by the engineering successes of the Victorian age and the twentieth century. Let's face it, the advances have been impressive. The discovery of antibiotics, for instance, has transformed our lives, reducing mortality once brought on by the most trivial infections. Things unimaginable two hundred years ago—like cooking your chicken with microwaves, mysterious and invisible electromagnetic radiation only discovered in 1888—have sometimes subtly and sometimes dramatically altered our lives.

Truly we have come to believe in our limitless capacity for technical ingenuity. As we progressed from horse and cart to car and plane, so we will one day advance to interstellar flight, as will other intelligences. But what if this is hubris? Might we one day confront something hitherto unknown to us—a technological ceiling? The nearest habitable planets that could host aliens might be hundreds or thousands of light-years away. As yet, we have no idea how to travel at an appreciable fraction of the speed of light, and even if we could, it would take many human lifetimes to reach those planets— planets that, on a cosmic scale, are our next door neighbors. There are just so many obstacles. Say we could muster the vast energy to propel a spaceship at even 10 percent of the speed of light, so that it would take a mere 830 years to reach Zeta^2 Cancri. Any spaceship traveling at such immense velocities would be ripped asunder by a collision with even the smallest grain of interstellar material.

No doubt, optimistic engineers are dreaming up ways of overcoming the challenges of speed and distance. Some think we might be able to break the ultimate physical barrier and travel faster than light. The idea is that we would send a ship into a wormhole, a distortion in the spacetime continuum that underlies the universe, allowing the ship to disappear in one place and reappear in another location of our choosing. This is purely a theoretical possibility; we haven't the faintest idea how to actually do it. Another fantastical prospect is to scrunch up spacetime in front of your spaceship, collapsing vast distances into short hops across the void. The Alcubierre drive, named for Miguel Alcubierre, the theoretical physicist who proposed it, is a device that might enable such a feat. Yet Alcubierre's ideas call upon extraordinary, speculative physics. We are not even sure if his physical theories describe our universe, let alone whether we could ever engineer the technology he has in mind. History cautions against rash dismissals of future technological possibilities—it was once thought that traveling faster than 40 kilometers per hour would likely kill a person. But the speed of light is not some arbitrary, self-imposed limit. Faster-than-light travel may yet turn out to be an insurmountable barrier.

That such a barrier should exist is hardly unreasonable. Physics itself is not limitless. Physical laws impose all manner of bounds on the stuff of the universe. Our understanding of those bounds is what enables us to build all our impressive gadgets, but it seems quite likely that physics also places limits on engineering. If we continue to expand our abilities, then at some point we will run into the edge. Faster-than-light travel might just be that limit. If it is, it applies to the aliens as much as it does to us. They too may be isolated in the numbing vastness of the universe. Like our own engineers, theirs stand impotent, chained by the finite possibilities of matter and energy.

One way around this problem is to take things slowly: accept the limits of physics and embrace the long delay between takeoff and landing. If you want to get to a star 10,000 light-years away by traveling at 1 percent of the speed of light—about 10,000 times faster than the speed of a jet airplane—doing so will take about a million years. This is a long time from the standpoint of an individual organism, but it is within the typical lifespan of a species on Earth. These numbers tell us that patience and persistence can get you far. But can any species endure such a journey? Can you really put thousands of individuals in an enclosed spaceship and send it into the black, cold nothingness of space for a million years and expect future generations to retain the sense of purpose that drove their absurdly distant forebears?

There is only so much isolation that humans can endure before our physiological and psychological limitations kick in. We know something of these limits thanks to studies of scientists who have spent time in the depths of Antarctica. They and their support staff have been examined with intensity by doctors and others keen to shed light on the parameters of human endurance. In the dark months of winter, a phalanx of psychological problems comes marching into view. Depression, loneliness, conflict, and outright derangement have all been observed alongside declining physical health. Immune systems become compromised under the pressure of separation, and the hormones scream stress. It is true that these populations are generally small, whereas we would be wise to cross the vast distances of interstellar space in worldships containing thousands of occupants. Doing so might prevent travelers declining into lonely insanity. Yet, given the fragility of the human mind and physique, we still cannot be sure that a crew of thousands, even tens of thousands, could stave off deterioration across ages and ages of travel.

Other potential solutions, such as genetic engineering and hormonal modification, come with their own problems. Say we could engineer humans to suppress all emotion, so that they don't succumb to existential dread while living their cloistered lives, whose lone preordained purpose is to bring about another generation that continues the voyage? Are these really the kinds of people we would want to send on this momentous trek through the stars? It might also be the case that a human devoid of emotions would be compromised in other ways that prevent carrying out the mission.

Yet even if such challenges could be overcome, we are still confronted by the question of why we or the aliens would make such a journey. An extreme emergency on the home planet might mandate migration across the forbidding cosmic ocean, but this would not be a voyage of exploration. The goal would not be first contact but rather a place to lay one's head. I suggested as much to my driver. Maybe, I told her, "they've got no motive to cross the huge distances of space. That would be a simple explanation."

My taxi driver seemed assuaged by this possibility. Perhaps the danger had receded. But now there was the prospect of loneliness. "I don't want them to be dangerous," she replied, "but being on our own with no one else to talk to. I don't want that either," she offered, with a tinge of sadness.

The silence we observe elsewhere in the galaxy can be explained with any number of hypotheses, but it's quite possible that the most obvious reason—they are not out there, or at least not nearby—is the answer. It is certainly worth our while to continue the search for intelligent alien life, but we may find that we come up short. If the aliens do turn up and are as bright as we hope, we might find some security in the idea that they have little reason to want to destroy us.

My taxi pulled up outside the entrance of Polaris House. I thanked my driver for the journey, but I had left her with this most human of conundrums. Be with others despite the uncertainty of the result, or persist in loneliness? Those are the choices that we, and the aliens, must live with.

This clay tablet, made in southern Iraq around 3,000 BCE, records information about workers' beer rations. Decoding alien language will be at least as challenging as understanding ancient writing, but we may be able to communicate with nonhuman intelligences on the basis of a shared capacity to understand science.

10

Will We Understand the Aliens?

A taxi ride to the University of Glasgow to borrow a Raman spectrometer and investigate samples that had been sent into space.

✳

My journey this day, a cold spring morning in 2017, featured not so much a conversation with a taxi driver as its opposite—a failure to communicate. Sometimes when you catch a taxi in Glasgow, you find yourself talking to a driver with a strong Scottish accent. It's a melodious, rich tone, but it can be difficult to understand when projected across a glass screen amid the rumble of a car engine and wheels, and especially when mixed with the ignorance of an English-born Edinburgh denizen like me.

My driver made what I thought was a comment about the weather. I caught a "nae" instead of a "no," the "g's had gone, and he pointed toward some menacing clouds that hung on the northern horizon. At times like this I feel a little rude because all I can do is nod, smile, and offer some weak affirmation of my interest. I suppose he was at least as cold as I was, though, wrapped as he was in a thick black woolen coat, his head popping up above a red scarf. If I was finding it hard to converse with my taxi driver, would I be any better off with aliens? The thought occurred to me that first contact could be a failure even if the aliens learned about Earth, including its languages, from a group of helpful Glaswegians or by watching Scottish TV from the comfort of their alien ship.

Then again, the barrier I confronted on this taxi ride was merely a linguistic one. If this could be overcome, my driver and I would have much to discuss. We might discover our disagreements but also our common outlooks. It seems obvious that this verbal barrier would exist with aliens. We would just have to find a way to communicate with them. But once the language problem was solved, would we share anything in common, as surely I did with my driver? Or would their alienness still render them utterly distant from us? Would it be possible to comprehend their mental state and their perspectives even if we established a common means of discourse?

As people so often like to point out, maybe the meeting of minds between a human and an alien would be akin to our own relationship with ants. Intellects far superior to ours would be no more able to extract a sensible conversation from us than you could from an ant, a bumblebee, or even a creature so advanced as a dog. That we possess intelligence far greater than dogs does not enable us to interpret their signals as effectively as other dogs can; something similar may be the case when it comes to aliens. Nor would it matter if it could be said that the aliens' intellectual capacity is roughly equal to our own. What matters is that an alien intelligence may be qualitatively distinct from human intelligence in a way that turns first contact into bemused silence.

And yet there is at least one dimension along which we and the aliens would probably be able to communicate: science. This is likely to be our common denominator. At the risk of sounding like an ancient philosopher defending the separation of man and beast on account of our power of reason, I am going to do exactly that, or something like it. The power to do science is a power of the human brain and not of other brains on Earth. I will not attempt a neuro-scientific explanation for this distinction, nor will I divert you into arguments about whether we are categorically different from chimpanzees or whether, in fact, all creatures lie along some continuum of cognition, with humans slightly more advanced than, but not definitively different from, our primate relatives. I simply want to observe that human beings build space telescopes and sit around with cups of tea discussing hypotheses about the beginning of the universe. Unless you are Gary Larson or spend most of your time in his imagined world, you probably agree that cows and monkeys don't do these things. And this makes all the difference in the world—or, I should say, all the difference in the universe.

But what does the human capacity for science have to do with our ability to communicate with alien arrivals? To make sense of this, we need to get a better handle on what is meant by "science," a word that is misused and mistreated in all sorts of ways. So let us begin by observing, perhaps surprisingly, that there is no such thing as science as such. You will often hear people say, "Science has shown . . ." Or, "Science can't explain everything." In the context of informal conversation, there is nothing egregious about these statements. Still, they misunderstand science in a profound way, pretending that it is a body of authoritative knowledge whereas in fact science is a method. A method is scientific to the extent that it involves gathering evidence from experiments or observations and then constructing on the basis of this evidence a picture of how nature works. That picture may be inaccurate or may contain contradictions, but the process of creating it was nonetheless a scientific one. And once you have your picture, you can use it to draw enlightened insights that enable hypotheses—conjectures based on your evidence. These hypotheses can themselves be tested through observation and experiment, and on and on as you grow your catalogue of information.

It's worth considering briefly how this process works. Say I have collected an apple and an orange and am studying their properties. I might, in a moment of effusive creativity, imagine that there exist fruits that are a mix of apples and oranges, a halfway apple-orange if you like. Let's call this proposed fruit an opple. Now I have a hypothesis and I can go out and test it by looking at lots of fruits in different orchards, searching for the enigmatic opple. At the end of the process, I'll either accept or reject my hypothesis—I'll either have an example of an opple, proving its existence, or I'll be faced with a suspicious lack of the fruit. This may not definitively prove the nonexistence of opples, but their absence from all accessible orchards should at least get me thinking that opples are extraordinarily rare. And, until some countervailing evidence arises, I will have strong reason to believe that there are no opples anywhere.

An unassailable principle of this exercise, and one followed rigorously by good scientists, is that you have to let go of your desires and prejudices and accept only what the data tell you, particularly if any information should definitively falsify your ideas. You might like to be the discoverer of the opple, with all the fame and tinsel that the eureka moment would bring. But were you to observe no such object, you would have to reject your hypothesis. It is not acceptable to pretend that you saw in a distant orchard an opple that has just, conveniently, died. You may not fake such fruits in your kitchen by means of some adept use of a paring knife or by other chicanery. Even if you've held on to your opple idea for a thousand years and are surrounded by a billion people who believe in the existence of the opple as vehemently as you do—if the data suggest otherwise, you must drop it.

That is science in a nutshell. It is not terribly complicated, yet it took an astonishingly long time to embed this simple process in the human mind. Millennia of superstition and religious dogma yielded other methods of understanding nature. The structure of the universe was in the tea leaves or was foretold by the entrails of a chicken. Most pervasive throughout time was, and remains, the argument from authority: things are as they are because someone powerful told me so. To the modern mind, it seems remarkable that no one seemed to think, "What a load of nonsense! I wonder how things really work—why don't I find out for myself?" But sometimes hindsight is easy. And many people did have just this suspicion; some even tried to act on it. But of course laboratories and accurate measuring tools did not exist in most places at most times, and support from elders was not necessarily forthcoming. Many advances eventually emerged from Europe, but for ages the continent was far behind. It was not until the seventeenth century that academies of science came into existence there, and luminaries such as Francis Bacon and Galileo Galilei laid the groundwork for the scientific method as we know it today.

I am certain that, when it comes to the scientific method, we and the aliens will be able to meet in the middle, which is to say that I am convinced the aliens also use science to enlighten themselves about the universe. Why such confidence? After all, people often say that science is only one way of understanding the nature of things, and other ways should not be dismissed. But while this statement has an appealing rhetorical resonance, and while it is trivially true, it rather misses the basic point that the scientific method is uniquely useful in adding to our knowledge of the universe. No one disputes that you could employ other methods; you can indeed consult a chicken's guts or peer into the bottom of your teapot or just ask someone from a special cult. But the question you must ask yourself is how reliable these approaches are. Will they bring you knowledge that holds up? Can you use this knowledge to test and test again until you converge on something useful? In other words, can you use the knowledge derived from the chicken guts or the cult elder to make testable predictions? If you can't, then you haven't actually learned about the physical operations of the universe in any systematic way.

This is another way of saying that science, unlike tea leaves and cult elders, is a process, not a font of potentially faulty wisdom. And it is a process with certain requirements. For one thing, it requires that we observe the phenomena we wish to understand. Unless our goal is to know how tea leaves wilt in hot water, peering into our cups doesn't fulfill that requirement. When we focus on the phenomenon we want to know more about, though, we are likely to gain relatively reliable information. Equally crucial, and equally absent from entrails and especially cult elders, is that matter of dumping your favorite idea when evidence calls it into question. Again, you don't have to do this in order to investigate the world around you. But you must if your investigations are to return reliable information. The scientific method is powerful because it is a never-ending circle of questioning and therefore refining our insights into the inner workings of nature. Other investigative approaches don't foster such refinement, making their findings less dependable than those produced over time using the scientific method.

Now, you could claim that your understanding of the universe can never be examined using the tools the scientific method demands. Once more, that is all well and good in the sense that no one will deny you the right to assert your perspective. But if your knowledge cannot be tested in any way that allows it to be subject to critique, does that not strike you as rather too convenient? We should be highly suspicious of any view of the universe that claims by its nature to be untestable.

This brings me back to the importance of clarifying what people, particularly scientists, actually mean when they make certain claims about what "science" knows. When you hear someone declare, "Science has shown," what they actually mean is, "This idea or observation, reached by the gathering and testing of data, has led us to this current understanding. We may decide this view is wrong if we later find data that challenge it." Such careful phrasing would make you terribly boring at dinner parties, so the shorthand is understandable if you want to keep your friends. But what's going on here is not just the fine parsing of some linguistic subtlety; the difference between the shorthand and the careful phrasing is crucial if we are to appreciate why science isn't "just another" way of understanding the universe. Science is a process of critical thought that demands unending confrontation among possible explanations of observed phenomena and rigorous inquiry into the quality of the observations themselves. No scientist worth their lab badge would deny that the power of the method lies in its unceasing course of checking and rechecking, founded on the understanding that there are no final answers—only deeper fathoms to explore. Chicken entrails don't cut it.

One of the proofs of the scientific method's reliability is that scientists don't just spin off new theories, any of which is equally useful. Other methods can do that. What makes science special is that, when we follow its methods, we can develop theories that allow us to make predictions and even to build things. When the predictions come true (every time) and the thing we build works (every time), we know the theory accurately describes the nature of the world around us. For example, the theories on how lift and drag work allow you to design an airplane that will fly through the sky. Sure, there may be a little trial and error; wheels and wings are not without complication. But the scientific method affords us enough certainty about the behavior of the material world that we can build things on the basis of first principles and improve from there.

None of this is to say that it is impossible to build things when we don't have the scientific method on our side. More haphazard approaches of trial and error do not inevitably fail, which is why there was technological advancement before the seventeenth century. However, the scientific method enormously accelerated technological development, particularly where deep understanding of nature's complexities was required to achieve success. Without the scientific method, it was possible to intuit our way to stable, if extremely inefficient and unsafe, shelter. It was possible to construct seafaring vessels and agricultural tools. But it was not possible to build a spaceship. Well, at any rate, a society without the scientific method would have real difficulty building a Moon lander. If you want to reject this assertion, then I have a simple challenge for you. Gather three teams of engineers with no knowledge of aerospace engineering. Provide one with a bowl of chicken entrails, the second with a priest from a respected religious order, and the last an aerospace-engineering textbook informed by studies following the scientific method. Ask them to build a lunar lander that they will be required to test. Report back.

Lest you think I have deviated from our alien encounter, these observations on the character of science bring me at once to my point. If aliens have built a spaceship and have made first contact, then I can guarantee you—without knowing anything about their world, their culture, or their brains—that they did not build that ship using information extracted from the entrails of a Xoggle beast or from the proclamations of High Priest Zinglebrod, Ruler of the Sixth World and Master of the Universe. They did it with the scientific method. If it turns out that Zinglebrod was involved, we can say that this creature uses the scientific method or had available a library, or alien equivalent, that contained information gathered through this process. And this convergence on a universal form of thinking might also suggest that the scientific method itself could be a basis for communicating with the aliens.

Now I have no hesitation in saying that Zinglebrod's knowledge may be different from ours. Indeed, it may be vastly, almost ineffably, superior to ours. To say that humanity and Zinglebrod can and have learned truths about nature by applying the scientific method is not to make assumptions about how that knowledge is contemplated or used. There need be no equivalence in our grasp of the universe, our technological capacities, and our material knowledge. But the gulf between ourselves and Zinglebrod is not the same as the gulf between a human and an ant, or even between a human and a more cognitively advanced chimpanzee. The difference between us and Zinglebrod is one of quantity, while the equivalence is one of quality. Both humanity and spacefaring aliens push our understanding of the universe toward more and more reliable insights by using evidence to test, sustain, and reject theories.

It might also be pointed out that Zinglebrod's capacity for implementing the scientific method could be different from our own. Maybe the alien is enormously better at performing mathematical computations in its brain. Perhaps the way it orders and accesses knowledge is different, even weird. But none of this changes the categorical fact that aliens will use the scientific method. Let me be stronger: they *must* use the scientific method if they want to grasp information about the universe that allows them to build a working spaceship.

At least some of this information about the universe will be familiar to us. That is due to another feature of the scientific method: it works the same way, and studies the same universe, regardless of who or what is applying it and irrespective of what planet the inquiring entity is on. I am not going to make the mistake of claiming that science can reach some ultimate, objective foundation of reality, if only because I want to deny my philosopher friends the opportunity to make mincemeat of me by questioning this very possibility. But there is no error in asserting that the scientific method does improve our knowledge frameworks so that, over time, we approximate better and better a full understanding of phenomena. Newton's ideas on gravitation built on earlier thoughts and were later refined and advanced upon by Einstein's expansive work on the spacetime continuum. Newton's ideas remain largely accurate and useful for predicting what happens when you throw a ball and you want to predict its trajectory as it falls to the ground, but Einstein's genius greatly improved our projections when we try to understand things at the cosmological scale. Other scientists have thought a lot about what Einstein's theories get right and wrong, sometimes asserting their wrongness only to find later that they were right all along. And sometimes we find areas of his thought that need improvement. And so on in an endless iterative process in the direction of deeper, more cogent truths.

I can therefore boldly assert that, if the aliens arrive in a spaceship, they too, at a minimum, grasp Newton's laws of motion. Of course, they are Babblezig's laws, not Newton's, but that is a triviality. It doesn't matter how strange the aliens' brains seem to us; they will have converged on the same understandings as we have. If they have not, they will not be able to plan a spacecraft trajectory or calculate the effect of Earth's gravity on their planned landing. Alien spaceship designers understand the laws of gravity.

An important caveat is that the universality of physical laws does not imply that the aliens possess exactly the same scientific insights or technological capacity that we do. It is interesting to ponder whether the apprehension of certain physical laws or technological accomplishments necessarily follows from others—whether there is a deterministic path of discovery. My own view is that there is a certain directionality in scientific understanding. It would be difficult for Einstein to consider the spacetime continuum without a grasp of Newtonian mechanics. So too, it is difficult to construct a reliable model of how a solar system works without these laws. It seems that our ability to grasp certain facts about our universe must be built on prior understandings. If the aliens arrive in a spaceship, at a minimum it is likely that they will have an equal understanding of the universe, and potentially they will have a much superior one. However, it is extremely unlikely they could arrive with a knowledge of their matter-antimatter propulsion engines, or whatever got them here, and then express admiring and stunned incredulity when presented with a copy of Newton's *Principia*.

The matter of whether spacefaring aliens would follow a technological pathway similar to humanity's is enormously fun to contemplate. When I am bored on a train from Edinburgh to London, I like to play a little thought game in my head. I try to imagine whether it would have been possible for our society to reach its current technological achievement while having bypassed some basic advancements of bygone days. For instance, could a society invent nuclear power without having invented the wheel? Indeed, could my taxi be replaced by something without wheels? Well of course, I could be on the back of a horse with my taxi driver periodically shouting at me in Glaswegian to hang on as we lurched over another hole on the track, the road helpfully lit by electricity from the nearest nuclear fission reactor. Goods could be moved around Glasgow using the techniques of the ancients, with boxes rolled along on logs, the rear logs continually transferred to the front with painstaking effort. Continuing this line of thinking, all steps in the generation of nuclear power—from the discovery of uranium and its properties to the development of the theory of fission and ultimately the building of a reactor—seem possible without the wheel.

However, whether any of this is intellectually likely is another matter. Surely a technician, staring at a uranium-enrichment centrifuge under construction, would think to themselves, "If I stuck the axle of that centrifuge onto the bottom of a box and replaced the centrifuge with a disc, I could pull the container along a surface without having to keep moving logs around. Eureka!" Many components of nuclear generation, such as turbines and water-cooling pumps, involve parts rotating on axles. One would think the existence of such components would encourage some thoughts on the usefulness of wheels.

I think it is fair to speculate, then, that it is not just knowledge that tends to be additive and path-determined. A degree of technology determinism is also probable, at least at the broad scale of major technological capacities. The aliens may need different things than we do, resulting in different priorities; perhaps they consume nutrients photosynthetically and so never bothered to invent toasters. But the electricity that powers toasters—that is something spacefaring aliens can be expected to understand very well indeed. Just as *Principia* will not stun alien visitors, it is also unlikely that they will land on Earth, gather round the wheel of a Volkswagen, and start muttering what our alien translation machines interpret as, "You've got to be kidding. Zog, take a look at this round thing. How did we never come up with this?"

If we meet aliens, the task of communicating may not be easy. We will be fortunate if they communicate using discernible noises or signs; their language structure and their means of processing information might be wildly foreign to anything we can imagine. Even their sensory perception might be quite different from ours. But I believe this will not be a meeting of ants and humans. We will look at each other and, through the mists of linguistic incomprehension, understand each other to be scientists. The capacity and desire to ask questions of the cosmos and to use observation, experiments, and critique to converge on some better grasp of our surroundings will make us equals, whatever inequality exists in the quantity and application of those capacities. It may even be that, as we survey their technology and they ours, the products of our quest to unravel the nature of the infinite void will bring us instantly into mental alignment, a sort of mutual respect and understanding of our common past and future as scientists.

The scientific method sets a species on a course for potentially endless advance in its insights into the universe. Although we know of no other species that thinks this way, there is no reason to believe that the scientific method is accessible only to humans. Far more, science is a necessary way of thinking if any species is to make systematic improvements in its grasp of the workings of nature. Whatever else may be different between ourselves and the aliens, we will have the luxury of entering into first contact with an unspoken understanding of these realities. We will understand something about each other. I, for one, would be fascinated to learn the alien word for "science."

NASA's eXtreme Deep Field image, assembled from ten years of Hubble Space Telescope photographs taken in a single part of the sky, shows about 5,500 galaxies. Could it be that no one is out there staring back at Earth, an infinitesimal speck in a different deep-field view of the universe?

11

Might the Universe Be Devoid of Aliens?

A taxi ride from Bruntsfield to Edinburgh New Town to attend a Christmas party.

✳

We turned onto Princes Street, and for the first time in the year I felt that ineffable emotion: Christmassy. We all know what melancholy, happiness, and envy are. They are the raw stuff of human experience. But Christmassy—what is that?

Actually, I think it's a complicated thing. Childhood memories, dark evenings, mulled wine, trees decorated with tinsel and assorted baubles. A whole lot goes into this emotional state, exaggerated by a collective seasonal hysteria. At the root of it all, though, is the sociability of the holiday, the sense of family and community.

"I've got the whole family coming for Christmas," my driver said. "The lot. Eleven of them all descending on me and my other half." All this came out of the blue, a sign of my driver's excitement. She looked ready. She wore a red-and-green jumper and even her hair, white as snow, seemed like it was in the mood for Christmas. "I'm looking forward to it," she added cheerfully, in case there was any doubt. "You too?"

I was, in fact, but as a space enthusiast, I can sometimes have a weird perspective on our earthly rituals. Here we have this little fleck of rock, covered in humans, some of whom celebrate Christmas. They enjoy each other's company, raise a glass, tuck into turkey, stash presents under a tree, all the while orbiting an unremarkable star in a rather lonely part of the galaxy. I try not to depress people with astronomical thoughts, so I wasn't going to bring up our infinitesimal meaninglessness. Nor was I going to ask my driver whether it mattered if there was anyone else in the universe or not, although such thoughts were on my mind. Would Christmas be better if we knew there were other creatures out there? Or might it be better if we discover that we are lonely, the warmth of our togetherness magnified by the lifelessness of the surrounding vacuum? Now that is one way to feel Christmassy: by embracing this flicker of color, gaiety, and hope in the blackness.

All that occurred to me in a flash. Then I caught up to the conversation. "Yes, I'm very much looking forward to it," I said. "I'm also seeing family this Christmas, and it's good to know we are not alone. At least not alone on Earth, but who knows about the rest of the universe." My driver said nothing. She looked into her mirror and squinted. Like I said, I'm a space enthusiast; I bait people to get an interesting dialogue going.

"You one of those *Star Trek* types?" she asked. I'm not particularly, although I do watch it occasionally. But before I could utter my firm denial, my driver cut in. "I'm really into *Star Trek*. It's all a big adventure traveling around and all that and meeting all those strange people."

Thinking back to my Christmas musings, I wondered whether the whole edifice of *Star Trek* would crumble if the mission to "seek out new life and new civilizations" were a perpetual failure. Did my driver enjoy the show precisely because it did involve talking aliens? Would a lonely *Star Trek* be as bad as a lonely Christmas? "I know this sounds a little strange," I warned her, "but do you think *Star Trek* would work at all if they didn't find any intelligent aliens to talk to. Maybe life, but nothing to chat to?"

"I like watching it to see who they meet," she replied. "What oddball characters they talk to and which ones try to mess up the ship." She tilted her head to one side and continued, "But I don't think it would work without them, would it?"

"I agree with you," I said, "at least that's what makes it fun to watch." From a viewer's perspective, it would be quite boring to spend an evening watching a ship travel though space, not up to much, even if it was going at warp speed. I'm pretty sure you'd agree, even if you don't watch *Star Trek*. But my affirmation of her views on the TV show was not in alignment with my thinking as a scientist. Professionally, I'd be ecstatic if I had the opportunity to join any journey across the cosmos, even if our crew didn't find a whole lot.

But please follow me through this rather dull fantasy for just a moment, for it is also instructive. Imagine an episode of *Star Trek* in which the starship *Enterprise,* on its five-year mission to explore strange new worlds, finds nothing. Or maybe Captain Kirk and the rest find microbes here and there but otherwise come up empty.

In the first year of the mission, boredom sets in. The craft leaps across the universe, warp-speeding from one dead solar system to another. By year three, Captain Kirk has taken to drugs and spends much of his time listening to albums by The Doors, while his languid crew sit around watching B-movies and daydreaming about the better jobs they might have in banking or real estate. By the conclusion of year five, they have visited well over 300 star systems, and all they have to show for it are geological samples and a few frozen vials of soil and ocean water, some of which seem to have something akin to bacteria in them. Kirk, bearded and disheveled, has almost lost the will to live, and the rest of the crew have become drunkards. They return to Earth, leave Starfleet, and take up jobs in an office block on the outskirts of Croydon, managing local road plans and overseeing pothole-remediation schemes.

Call this *Star Trek: The Documentary*. It may not be much fun, but it might be more lifelike than the real thing. *Star Trek* as it actually exists reflects society's past optimism about aliens. As you will recall, for centuries people thought Mars and Venus were abodes of intelligent creatures, civilizations going about their daily lives just as humans do. The Moon, that gray, baked wasteland, was the home of Lunarians. Strange lines on Mars, mistaken for canals, were the artifice of aliens, ambitious engineering projects that bent the Martian environment to the inhabitants' will. These observations, propelled by such luminaries as Huygens, Herschel, and Lowell, stimulated the public mind and provoked not just confidence in the existence of alien intelligence but virtual certainty.

What changed all this was the space age. First, the earliest high-quality images of Venus, Mars, and the Moon showed pretty clearly that these worlds were nothing but rock. Later, more advanced studies eliminated the final shreds of hope for our neighbors, proving definitively that there are no other civilizations in our solar system. Yet the question of whether there is life elsewhere remains open and fetching. Hence the continuing appeal of many science fiction stories.

And of course the search for life elsewhere remains appealing for scientists, too, but our work looks more like that depicted in *Star Trek: The Documentary* (minus the substance abuse, one hopes) than anything to be found in the TV series and movies. One of our favorite research subjects is the surface of Mars, which contains extensive evidence of ancient bodies of water: primordial minerals and clays that were formed in water, braided channels that were once tributaries of rivers, and fan-shaped deltas bearing witness to lakes that existed at a time when the Martian atmosphere was thicker than it is today and liquid water was stable on the planet. Today there is ice on Mars, but when heated the ice instantly vaporizes into wisps of gas, bypassing the liquid phase altogether. If there was ever life on Mars, and if there is today, it was and is likely only microbial. Certainly there is no indication that complex animal-like life ever roamed its surface.

Beyond Mars, our probes have discovered oceans under the icy crusts of moons orbiting the gas giants Jupiter and Saturn, igniting interest in the possibility of life there. Jupiter's moon Europa is no larger than Earth's satellite, yet it may have twice as much water as all the oceans on Earth combined. Saturn's moon Enceladus is even more unassuming than Europa—merely 500 kilometers across, less than the length of the United Kingdom. But it too deserves special attention. Enceladus spews into space diffuse jets of water that contain organic material, hydrogen, and a cocktail of other ingredients that tell us its underground ocean may be tolerable for life.

Were scientists to discover microbes on these watery worlds, we would be overjoyed. Yet the public may be disappointed, for extraterrestrial microbes are not necessarily alien. That is because the assorted rocks strewn about the solar system have been sharing material since time immemorial. When an asteroid or comet collides with a planet or other object, the impact ejects rocks from the surface into the far reaches of space. Lots of rocks—not pebbles but mountains upon mountains. These violent impacts are infrequent, but the amount of material they hurl into space should not be underestimated. These days, about half a ton of Mars penetrates Earth's atmosphere in a given year and comes crashing to the surface. If you are not the lucky recipient of chunks of Mars rock, it is because most of this apparently formidable amount of material lands in the oceans or in unoccupied deserts. The chance of a piece of Mars landing in your garden is small indeed.

In the course of geologic time, though, cosmic objects do exchange a fair amount of their contents. And within these chunks of rock, microbes might survive. Scientists have simulated impact conditions by accelerating small lumps of bacteria-soaked rock to high speed and slamming them into solid targets, and we have found that the bugs can withstand the intense shock pressures. It is therefore possible that microbes from Earth have arrived on Mars, and vice versa. In fact, as outlandish as this sounds, some people have suggested that earthly life originated on Mars and was transferred here. Maybe we and everything living on Earth are Martians. That would be a poetic and ironic conclusion to draw from our Mars explorations.

The possibility that the planets have been exchanging life, if indeed it is to be found elsewhere, brings us to the perhaps-disheartening realization that any life found elsewhere in our own star system may be just like life on Earth, or else clearly related. That would not make this foreign life uninteresting; researchers in fields from psychology to sociology to genetics learn a great deal from twins separated at birth, and, likewise, we could gain considerable knowledge by studying what our cousins have been doing for the past few billion years. But the microbes would not be entirely alien. Their origins and trajectory would be tied up with our own. For otherness, true alien credentials, we might hope that we not only find life somewhere else but that it followed an arc independent from that of Earth life. Then we would have genuine, bona fide aliens to examine.

All this is a far cry from *Star Trek*. The crew never beamed down to the surface of a planet, collected some microbes, and spent the rest of the episode examining them under microscopes while engaged in long discussions of microbial ecology. The only microbes *Star Trek*'s writers deem interesting are those that interfere with the *Enterprise* and its crew in a way that suggests self-awareness. I have to admit some disappointment. I personally think that a *Star Trek* involving the study of microbial life across the universe would be fascinating and educational, but you are reading the words of a microbiologist, and I suspect you may not agree. My driver didn't.

"Do you think *Star Trek* would work with at least some life though?" I asked. "Not pointy-eared sentience, but lots of interesting microbes and other strange creatures in the rocks and soils. With a bit of work, do you think that could be interesting?" I knew as soon as I had broached that question that I was showing my credentials as a card-carrying geek. I mean really, could anyone expect the public to get excited by thirty or forty minutes of lab experiments, even if they were carried out on the *Enterprise*? My driver didn't budge.

"Well there wouldn't be much to watch, would there?" she said, as we turned onto George Street. The long rows of eighteenth-century façades were decorated with greens, reds, and silvers, lanterns, lights, and cheer.

The *Enterprise* was not confined to the solar system. In journeys to the far reaches of the galaxy, might we meet with better luck? At the moment, we do not know. We are working on it, though. One of the most auspicious developments in astronomy over the last three decades has been the hunt for planets, similar to Earth, orbiting other stars. These so-called exoplanets have already revolutionized how we think about the cosmos and what we seek in it. So far, telescopes such as NASA's Kepler and TESS (Transiting Exoplanet Survey Satellite) have shown that an immense variety of these exoplanets exist and that some of them orbit parent stars at a distance appropriate for the existence of liquid water. Such planets occupy what is known as the habitable zone, the ring around a star in which the planet's surface receives just the right amount of stellar radiation: neither so much that the planet overheats and water boils, nor so little that water freezes solid. Furthermore, many of the planets discovered in the "Goldilocks" zone of one or another star are rocky, as opposed to gaseous, and therefore potentially suitable for life. A real-life Kirk would have plenty of places to visit.

In the next two decades, increasingly powerful telescopes will allow scientists to look at the gases contained in the atmospheres of these distant worlds, giving us further information about their suitability for life. One might wonder how this is possible, given that telescopes observe light from afar; we won't be sending probes to take chemical samples of exoplanet atmospheres. Spectroscopy—this technique of determining the composition of matter on the basis of the light it emits, reflects, or absorbs—is nothing new, though. In the case of exoplanets, what interests us is the light our scopes won't see. As starlight passes through a planet's atmosphere, the gases within it absorb certain wavelengths of the light. These missing wavelengths show up in our detectors as troughs in the intensity of light in the relevant part of the spectrum. In turn, these troughs are fingerprints of particular gases. For instance, if the light entering our telescope is missing particular wavelengths associated with oxygen, then we know that oxygen is present in the atmosphere. In this way, scientists can figure out what an exoplanet's atmosphere is made of just by scanning the light passing through it.

Many of the gases we will find are those we would expect to find in the atmosphere of any rocky planet, such as carbon dioxide and nitrogen. If we are lucky, our instruments might also spy the telltale signatures of water. A planet that has lots of water in its atmosphere would be very exciting because it probably also has large amounts of water on its surface, perhaps oceans conducive to life.

Gases indicative of habitability are critical, but we need not stop there. We can also look for gases that suggest the presence of life itself. To find life, we need to look for gases that living things produce. That is a tall order because many gases that are the byproducts of living processes can also be produced by geological processes, so they are not a surefire signature. Nonetheless, some gases have promise. Oxygen is a product of photosynthesis, so if we find oxygen in an exoplanet's atmosphere, that is a strong sign of biology on the surface. The remarkably high oxygen content of our own atmosphere—21 percent oxygen—is the accumulated waste product of bacteria, algae, and plants churning through carbon dioxide, absorbing sunlight, and making the sugars they need to grow. Were a similar proportion of oxygen found in an exoplanet's atmosphere, the scientific community might respond with considerable glee.

Might. Unfortunately, even prodigious oxygen is not a guarantee of life. Abundant oxygen can be produced in the absence of any organic processes: break up enough water with intense radiation, and it will turn into its two constituent materials, hydrogen and oxygen. However, with forethought and careful use of computer models, we can work out exactly when we would expect to see false-positive detections of life on the basis of oxygen-rich atmospheres, allowing us to rule out candidates before giddiness sets in.

The problem for Kirk is that, even if we do find oxygen on a habitable planet, this does not imply the presence of intelligent life. Certainly oxygen is necessary for an intelligence like ourselves to gather energy from its environment. But a life-bearing planet with oxygen could be a soup of bacteria bubbling forth the gas without a Klingon in sight. We should therefore be prepared for the possibility that we will eventually find life visible only with a microscope. The universe may be dominated by simple creatures.

Should the possibility of a universe devoid of bounteous civilizations disappoint us? Whether or not it should, there can be no doubt that it will. I would be let down, and maybe you would as well. This is an intensely human reaction. We want to know we are not alone, and we yearn for the thrill of joining an intergalactic community. There is much to be said for a future of diverse, never-ending, and thought-provoking conversations with otherworldly minds. There is nothing wrong with this sense of anticipation, which impels us to continue exploring and to do what the *Star Trek* mission itself promises—to seek out strange new worlds and new civilizations.

Of course, it would be difficult, in fact impossible, to show that there is no alien life, intelligent or otherwise, anywhere else in the universe. How can we ever know that there is not one isolated society on a planet billions of light-years away? But say we searched thousands of earthlike worlds that had all the ingredients of life, the most propitious candidates we could find in our galactic neighborhood, and all of them were barren. What might this tell us?

Well, besides the obvious conclusion that intelligent civilizations are rare, we could investigate these planets to find out whether they do, or did, host any life at all, even microbial life. We may find we are bereft of aliens to talk to, but also that the universe is full of bacteria-like entities. That would be important because it would tell us that life can easily start, but its journey from simple replicating cell to advanced forms and ultimately intelligence is unusual. Something along that path is difficult to achieve.

Or we might instead find that we live in a universe of planets that have all the ingredients for life, but nearly all of them are sterile. This outcome—a universe replete with habitable but lifeless worlds—would be astonishing and informative in itself, demonstrating that the conditions that allow for life are common, but the concatenation of events that turn chemical compounds into replicating, evolving life are rare. In this case, it could be that intelligences easily emerge from microbes, but microbes themselves are hard to create—that the emergence of life is an extremely delicate process demanding conditions that almost never prevail.

There are many different ways for a universe to be quiet, many more ways than there are for a universe to contain intelligent creatures. Every one of these scenarios has much to tell us about our own origin, how likely it was, and what pitfalls and chance events might have forestalled our emergence. Was the origin of life on Earth a near-miraculous event? Was the ascent of complex multicellular life unusual? Were the conditions that gave rise to intelligence particular?

To grapple with these questions in a meaningful way, we need to seek out life and intelligence on other worlds. Only then can we achieve the sort of knowledge that enables firm conclusions. This is to say that a lot can be gained even if Captain Kirk never goes on an obsessive quest for civilizations across the universe. His scientific portfolio would be much enriched if he also studied lifeless worlds and planets with primitive organisms, for then the *Enterprise* would truly be on a quest to understand life in the universe. These aspirations might seem a little uninspired to the *Star Trek* audience, but I'm pretty sure that Spock, that logically minded Vulcan, would concur. Science doesn't seek to fulfill fantasies and wishes. Its job is to test hypotheses in an effort to give us some grasp on how our universe works.

Call me a bore, but I've always thought the crew of the *Enterprise* was doing bad science. The opening lines should have been, "Space, the final frontier. These are the voyages of the starship *Enterprise*. Its five-year mission: to explore strange new worlds, to test the hypothesis of alien life and understand the factors that give rise to microbial or uninhabited worlds, to boldly go where no one has gone before." But I suspect I would have been fired as a scriptwriter.

To find new civilizations in the universe would be a stupendous thing, and let's not be dull in discouraging the human imagination. But let's also remember that whatever we do, or in particular don't, find will tell us much about ourselves and our place in the universe. The confirmation of a quiet and lonely universe would greatly expand our understanding about our place within it. If Kirk and his crew return empty-handed, their five-year mission will still have been a great success.

The Martian surface is an extreme environment in many ways. This composite photo taken by the NASA *Curiosity* rover shows the vehicle's wheel tracks in the dry, radiation-soaked sands of the Red Planet.

12

Is Mars an Awful Place to Live?

A taxi ride to Boulby Mine in Yorkshire to oversee a test of planetary exploration rovers in our underground laboratory.

The conversation got interesting twenty minutes into a drive across the Yorkshire Moors. Now don't get me wrong, the moors are stunning, but in the middle of nowhere it can be nice to pass the time talking.

"It's beautiful around here," I said, "but it's amazing how quickly you are in the countryside. I mean if your car broke down, it would be a real pain out here."

My driver nodded. "You're not wrong there," he laughed. He was middle-aged, and his accent was more southern England than Yorkshire. He sported a blue shirt and a pair of blue-rimmed glasses. His fingers tapped the outside side of the window frame while his arm rested on the open window.

I was heading to Boulby Mine, a thousand kilometers of labyrinthine roadways nearly a mile underground, where my colleagues and I had spent several years testing rovers and other space-exploration technologies. It was our little piece of Mars deep under Yorkshire, if you like. For some time, the mine has hosted one of the most impressive underground science laboratories in the world. Ensconced in its quarter-of-a-billion-year-old salty tunnels, the ultraclean air-conditioned laboratory is like something out of a science fiction movie. Here, scientists search for the elusive dark matter, one component of what we think the universe is made of. And down in those tunnels live microbes that slowly munch away on the ancient food in the salt. They have learned to live in perpetual darkness. While cosmologists take advantage of the tunnel's depth to block out radiation and stray particles that could contaminate their dark matter–seeking instruments, we students of life try to find out what the microbes are up to.

Ancient salts have also been found on Mars—salts that can wreak havoc on our cameras, environmental monitors, and other instruments. So it makes sense to test our designs in places like Boulby, to make sure they hold up. In reverse, instruments that are small, lightweight, rugged, and ready for space can be used to improve mining on Earth, perhaps allowing us to do it more cleanly or get better use out of the planet's scarce resources. In other words, deep in the tunnels of Boulby, space exploration and an Earth-based challenge—successful and sustainable mining—come face to face. This effort has brought to Yorkshire teams from NASA and the European Space Agency and from India, where the rovers being tested here were built. We have even hosted an astronaut who did part of his training in the mine, digging and scraping away at briny extrusions to learn how to collect samples in future planetary missions. It is a thrilling and exciting way to indulge interests in space while solving problems here on the home planet.

"It is a lovely place," my driver agreed, scanning the moors in front of us, "but it can feel like another world." He had made the terrible mistake, so common among taxi drivers, of offering me an inroad to talking about Mars. For an astrobiologist, the phrase "another world" is like a red rag to a bull. I made my move.

"Talking of other worlds," I said, "would you actually go to another world. Like Mars?"

"It's cold, isn't it?" he replied. "Much colder than Yorkshire. Can't say I'd leap at the chance, but maybe. People have gone everywhere, so I think we'll go and make it a home. Maybe one day they'll have a city. But it still won't be Yorkshire." It was the last comment that struck me most forcefully, the casual certainty of that comparison.

"A home on Mars," I said. "Would you do it?"

"Absolutely not," he emphasized. "Those space billionaires, they are welcome to it. It's far too extreme and I like Yorkshire."

The uncomplicated finality of a view like this can be disappointing for someone like me who gets excited about Mars. It's like finding out that your dinner companion shares none of your interests and any conversation about Mars and space is off the table. But there was more to my driver's perspective than a mere disinterest in Mars. As we drove through the beauty of the scenery, it occurred to me that Yorkshire was just where my driver wanted to be. Whatever he thought of Mars, he could be entirely satisfied with his place on Earth. Who could blame him, with the moors all around us? He was home.

A home on Mars. Those four words conjure up exotic thoughts of spaceships, futuristic spacesuits, even exploration attire for the family dog. Countless generations have dreamed of new lives on the frontier of Mars. These are would-be extraterrestrial pilgrims. Life will be hard at the start, but as more people flock to the bright uplands of the Red Planet, the day to day will become easier. And hey, you were the first, the wave of immigrants that founded it all. Now who wouldn't take that opportunity to found a new world, to establish a branch of our civilization on a faraway shore? A twenty-first-century rerun of the earliest settlement of the Americas, but this time without displacement, exploitation, and destruction, for there is no intelligent life on Mars. This will be a morally clean frontier, a manifest destiny the whole species can be proud of.

As in the Wild West, living off the land will take some ingenuity, but we do know how to do it. For instance, fuel can be made from the Martian atmosphere, but to do it you have to throw away your prejudices about hydrocarbons. Not that there is oil to be drilled from the Martian underground; as far as we know, Mars had no ancient biosphere that could in time yield fossil fuels. However, the Martian atmosphere does contain a lot of carbon dioxide gas, and that's a start. Mix it with some hydrogen, which you can get by splitting water with electricity (the electricity could come from wind or nuclear power to begin with), and gently heat your mixture over a metal catalyst. What drops out of the other side is methane. Liquefy this gas, mix it with some oxygen that you can also gather from carbon dioxide, and you have something you can burn in a stove or use to top up your rover and take a journey across the Red Planet.

There is something bucolic about this existence. The family sitting round the methane stove on a cold Martian winter's day. The faint wind humming as it whips round the edges of the habitat. The children readying themselves for an encounter with the extremes of the environment in their pressurized wagon. Reminiscing about Earth and times past.

It's a little like Charles Portis's 1968 novel *True Grit*, albeit set against the gleaming silver technological outlines of a Martian settlement. It's easy to see how this vision of living off the land and confronting the extremities of the Martian frontier mutates into a heroic fantasy that appeals to anyone who aspires for a place in history or a rewind to a tougher and more vigorous existence. So it is not surprising that the Martian frontier is held up by many people in the space-exploration fraternity as a place where humanity's dreams of expansion can be realized. Here is a theater in which to play our noblest selves.

There is undeniably some truth in the dream—not least in the sense that we will need to be at our best if we are to succeed in making a home of Mars. Settlement will be difficult; the challenge will likely claim some lives, and it will certainly stretch human ingenuity to its limits. Mars will demand everything we have. It will relentlessly grind away at our resolve, forcing us to new heights of determination and resilience.

Mars in fact is so unforgiving, and Yorkshire so lovely, that one might be readily forgiven for siding with my taxi driver. For a start, the Martian atmosphere is 95 percent carbon dioxide and only 0.14 percent oxygen, a combination lethal to humans. Not only that, but the atmospheric pressure is about one one-hundredth of what we experience on Earth. To all intents and purposes, the Martian atmosphere might as well be a vacuum with traces of poison. Our settlers will need a spacesuit to go outside, and their houses must be hermetically sealed so that they can be pressurized and filled with breathable gas including oxygen, the substance of life.

This difference from Earth, this noxious blanket of gas, makes all the difference to our Wild West notion of Mars. The restrictions on movement caused by Mars's asphyxiating atmosphere are worse than any restrictions faced by the Europeans who colonized the New World. Sure, they needed to be aware of rattlesnakes, flash floods, and indigenous communities defending themselves from the colonists' unfriendly intentions. But none of these hazards were omnipresent and all could be mitigated more or less easily, whereas the Martian atmosphere is both pervasive and will kill the unprepared in seconds. There is no reprieve; it stalks you wherever you go, eating at the sense of freedom, to say nothing of safety. One crack in a visor, a leak in a habitat, and all is lost. This is an extreme that no settler has ever experienced on Earth.

Unfortunately, it gets worse, for if the minimal, toxic atmosphere doesn't kill you, the sheer desolation might. It wasn't always this way. More than 3 billion years ago, Mars was covered in lakes and rivers. Orbiting spacecraft have taken detailed photos of meandering geological snakes, sinuous and undulating depressions that wind their way across the bleak landscape. In the northern hemisphere, there may even have been an ocean. Then the planet cooled, and everything changed. Now, ages ago, Earth cooled as well, its volcanic adolescence giving way to the more temperate adulthood we rely on. But this process happened more slowly on Earth than on Mars, which has a lot to do with why today's Earth, unlike Mars, has a life-maintaining atmosphere and a vast ecology. The Red Planet is about half the diameter of our own, and just as a bun cools more rapidly than does a large loaf of bread, Mars dissipated its heat more quickly than Earth did—so quickly that Mars's molten core stopped churning, in turn curbing the planet's magnetic dynamo and so its magnetic field. As a result, Mars was no longer shielded from incoming streams of solar particles; Earth is similarly bombarded, but because its core kept agitated, it has a magnetic field to deflect most of those particles. On Mars, the solar radiation tore the unprotected atmosphere to shreds, leaving wisps of it to sputter outward into the void. Asteroid and comet impacts also did their bit to heat the gases, leading to more diffusion. As the atmosphere thinned, so the pressure on the surface became too low to maintain liquid water. Instead, it froze into the ground.

What remained were dried and exposed rocks. The relentless wind has been grinding them down for millennia, leaving tiny fragments strewn across the Martian surface, which give the planet its everywhere-red complexion. Today Mars is an expanse of desert covered in ochre dust. No Sahara, Mohave, or Namib can compare to Mars's all-encompassing aridity. In those eroded rocks may lie an answer as to whether, in Mars's more fluvial past, it was inhabited by creatures making a living in its rivers and lakes. Maybe some vestige of that life exists underground on Mars today. It is this enthralling possibility that tantalizes scientists and explorers, inspiring them to someday make the journey to our nearest planetary neighbor.

However, whether Mars hosted microbes, and whether they still chomp at the rocks under the surface, means little for our merry band of pioneers sitting around the methane fire. For them, the consequences of this Martian history are all too clear. There are no rivers, no lakes, not even a bubbling little spring in which they might dip their cupped hands for a sip of water. There is no plant life. Not even a dried and shriveled tumbleweed rolls across the Martian desert. Mars is deader than the most lethal desert on Earth. Even in the most remote reaches of the Sahara, faced with imminent demise from starvation and thirst, at least you can face your maker with a breath of air.

As the taxi weaved in and out of small lanes, the moors were momentarily interrupted by human habitation. To my right a village shop appeared around a corner. Three people sat outside, and an old red phone box stood as a vigil to a bygone era. I wondered whether my driver was up for thinking about Mars as a new home for humanity generally, even if he wasn't too enthusiastic about it himself. "I don't want to press the point," I began, "but do you think Mars will ever be a new frontier? Do you think it will be a second home for anyone? I agree its not the moors, but . . ."

"Why not?" he said. "We've gone everywhere else. Once we decide to do something, we go, so I think people will go. Didn't they have people volunteer to go and live there? Yes, yes, like a second home. Well it is really, isn't it? But it's still very cold. I still prefer Yorkshire."

He was a little more upbeat in his assessment this time round, but the moors still trumped Mars, and I understand why. Mars beckons, but is there any sensible comparison with the purples, greens, and pinks of the moors, the sundews, the heather, and the cranberries? What of the plovers and merlins, the cuckoos and curlews that lunge about this windswept oasis in the north of England? My mind drifted back and forth from the moors to Mars, from Mars to the moors.

Yorkshire is no warm paradise, but as my driver knew, Mars compares badly here as well. Unprotected by the thicker greenhouse-like atmosphere that keeps Earth's temperature comfortably above the numbing coldness of space, Mars must endure forbidding extremes. At the equator, under the full glare of the sun, temperatures can be pleasant, exceeding 20°C. But across most of the planet, conditions are frigid. The average temperature on Mars is about –60°C, and the polar ice caps are chilly even by the standards of Antarctica—a cool –150°C.

The residents of the bleak, cold Martian desert will face still another enemy: solar radiation. With no oxygen to speak of, Mars lacks that protective ozone shield that screens out much of the sun's ultraviolet rays. On the surface of Mars, you'd get a tan about a thousand times faster than on Earth.

Of course, you'd be breathless and dead before you could enjoy your beetroot sunburn. More importantly from the perspective of a future Mars settlement, the radiation level at the surface will wither and yellow plastics, weaken other materials, and kill unshielded crops.

Silent invaders from the sun and the galaxy join in for the ride. Protons and high-energy ions stream onto the Martian surface from our sun and from elsewhere. Earth's surface is guarded by our atmosphere and magnetic field, so we experience about a hundredth of the onslaught that Mars does. A resident of the Red Planet would be at high risk of cancers and radiation damage. Slowly but inexorably, the Martian environment eats away at DNA.

Lest you think I exaggerate these matters, let's be clear that Mars lacks some of the dangers the Jamestown settlers faced. There are no indigenous Martians, weapons in hand, waiting to launch a nighttime attack. Our Martian settlers can sleep safe at night. Mars also features no ferocious storms or hurricanes that could wipe away the settlers' crops or shatter their homes. Mars is also seismically relatively inert, so its residents will never be harassed by volcanoes or earthquakes. This is not to say that Martian nature is easygoing, obviously. In addition to the many hazards we have already encountered, dust circles the planet perpetually, throwing red grit into every nook and cranny.

Alongside physical dangers, Mars surely poses challenges to mental health. Stare out of your Martian habitat, and as far as the eye can see there is red, some orange, then more red woven into a salmon-colored sky. Kick the soil and you might see the gray, unweathered basalt rocks that lie beneath the red dust. But Earth's blue sky; the green of the trees; the pinks, blues, oranges, purples, and fawns of the springtime blossoms—for settlers on Mars, these colors are confined to a computer screen, a plant-growth unit, a lone bud eking out a life on a habitat windowsill. Otherwise there is endless red. Can you endure this monotony?

There is little doubt that this dead planet holds great promise for scientists. For those who want to discover whether Mars ever hosted life, a playground awaits. Most of Mars's rocks are more than 3 billion years old, which means that traces of its potentially habitable past could be there for the finding. On Earth, by contrast, most of the ancient rocks have long since been destroyed by plate tectonics, the grinding of the continents that ceaselessly buries, pressurizes, and heats rocks into oblivion. Mars is unencumbered by this process, making it a window to the geological, and maybe biological, processes of the early planets. And if Mars is found to be devoid of life, fascinating questions emerge. Why, despite the presence of rocks and water, did this planet remain sterile, while its earthly sister sprouted and blossomed? Scientists of many persuasions, myself included, would grasp with alacrity a ticket to Mars.

Perhaps tourists would want to experience this desert too. They trek across the Sahara, they drive 4×4s into Death Valley, they mount camels to cross the Namib. Surely they would hop into the back of a pressurized rover to drive across the great Elysium plains of Mars, to peer into the

5-kilometer-deep chasm of the Valles Marineris, or to stand and stare across the vast white wilderness that is the Martian north polar ice cap. Yes, one can easily imagine that an adventurous soul would put Mars on their list of dream getaways. But just as easily, one can imagine that after a week or two, they would be ready to head back to Earth. Few Sahara trekkers spontaneously decide to abandon the comforts of home and remain in the desert for the rest of their lives. Similarly Mars may be suitable as a rare, expensive, and novel experience but not as a home.

Matters become less predictable when considering Mars's exploitation for economic purposes. The prospect of financial success drives people to do things that others find crazy. What if something of value is buried beneath the Martian dust? We don't know of any mineral or rare ore on Mars that would lure a rush of miners, but who knows. It is not inconceivable. Mars could also be a staging ground for other activities in the solar system, such as mining the vast quantities of platinum-group elements, iron, and water in the asteroid belt that lies between the Red Planet and Jupiter. Compared to other options in that region of the solar system, Mars is relatively well protected. A mining concern might house workers and equipment there.

Even then, it is hard to imagine putting down roots in the rocky desert. Short-term habitation still seems most probable. We might conjure up a picture of a Mars station where a clutch of scientists, tourists, and miners huddle in a bar, brought together by a patchwork of interests. They will enjoy the camaraderie wrought by isolation. But they won't stay longer than they need to. The tourists will take their flight home, the scientists will stay until their funding dries up, and the miners will depart when their rotations end. Since no one has yet been to Mars, its sands remain shrouded in romance. But once humans experience that environment firsthand, minds might change.

We need only look to our own planet to be skeptical of a potential Martian real estate boom. Consider that the population density of the Canadian High Arctic is about 0.02 people per square kilometer. Compare that to London, with about 5,700 people per square kilometer. Why such a difference? Well, one might invoke reasons including the challenge of reaching the Arctic. Maybe the Canadian government just needs to be encouraging—give migrants citizenship and financial incentives and pay for their relocation. But I suspect that even this would not be enough. One could put in place all the logistics and amenities, and yet overwhelmingly people would prefer to stay where they are—or pay outrageous rents to enjoy the benefits of London—than suffer the −40°C winters and barren landscapes of the Canadian far north.

For some people, like the Inuit, these extremes are the stuff of home. But most of humanity, a fundamentally subtropical species, could never be paid enough to move to Nunavut, much less Mars—and Nunavut is a good deal more agreeable. The High Arctic has a breathable atmosphere. There is plenty of wildlife to lift the spirits. Water is relatively easy to access, and the radiation levels are much the same as anywhere else on Earth. Even the cold polar desert of the most extreme Arctic north offers more sensory variety than Mars. And while conditions are biting, they are not instantly lethal.

If we were to forget all those facts and assume for a moment that Mars was as appealing as the High Arctic and that we could get people there as readily as we can to the far north of Earth, and if we further assume that the result was a similar population density across the entirety of the Red Planet, then the Martian population would still be less than 3 million. That is a mere 0.04 percent of the population of Earth. This would certainly be a new outpost for humanity, a new string to our bow, and a profound reality with which to reckon. We would be a multiplanet species, as discussed in chapter 7. But Mars would hardly compare to Earth as a crucible for our civilization.

I suspect that many people will be taken by the dream of the Red Planet and will move to Mars when that eventually becomes possible. But I wonder how many will stay. Once the excitement of novelty wears off, how many will look on its dusty, boulder-strewn plains and yearn for the sound of birds, the splash of rain, the colors of autumn, and the green buds of spring? The hopeful and the restless may lay their heads on Mars awhile, but who among them will want to call it home?

As a scientist, I cannot help but be enthralled by images of Mars, its landscapes, and its water-world past. I want to learn everything I can from Mars. Yet I cannot shake the impression that in the far future many a person who ventures there will have the same experience Captain Robert Falcon Scott did when he arrived at the South Pole after two and a half months dragging sledges across a white wilderness. "Great God, this is an awful place," he exclaimed. Similar words will be muttered from Martian habitats someday. And I suspect that there will be a few who follow that up with, "Take me back to Yorkshire."

Lunar station designs, like this one from NASA, look exciting and futuristic.
But space settlers will be a dependent group, bound to airtight habitats,
machine-generated oxygen, and other life-support and safety systems that can
never be allowed to fail. How much freedom is there inside this station?

13

Will Space Be Full of Tyrannies or Free Societies?

A taxi ride from Waverley to Bruntsfield Avenue after a meeting about a scientific paper.

✦

"What do you do?" my driver asked as we turned up Market Street. He was fidgety, fiddling with papers on his dashboard and wriggling in his seat, perpetually searching for a comfortable position. He had short, wiry brown hair and wore a baggy oversized black T-shirt. He kept looking in the rearview mirror as if expecting conversation.

I was very tired and not that interested in talking, so I explained a little about my job and then tried to turn the conversation his way.

"Would you go into space if you had the chance?" I asked.

"I think I would," he replied. "It would be a real escape from here. For a while anyway. Maybe not to live."

I've already shared my views on otherworldly escape from a damaged Earth, so there is no need to dwell on my objections to this notion. But my driver seemed to have in mind less a departure from an Earth ruined by its stewards than an opportunity to try, at least for a while, a new way of life elsewhere. No doubt, this idea that space societies differ dramatically from earthly ones is commonplace. After all, Hollywood science fiction has long indulged our yearning for escape. Films like *Star Wars* and *Avatar* transform the universe into a playground for our fantasies, an endless expanse in which the imagination runs free. The worlds of this universe, being products of our own creation, reflect our hopes and fears. They may be any kind of utopia or else benighted by evil. Occasionally a talented writer dreams up a cosmic civilization whose complexity and depth rival humanity's. These entertainments raise an interesting question. What sort of societies might really emerge up there, and what would it be like to live in one of them? I thought I'd sound my driver out.

"But do you think you would escape anything?" I asked. "It's very extreme out there, and you'd depend on a lot of other people to survive."

"I know you'd be in a tin can, but you'd still get away from all the trouble here," he insisted.

"But maybe to end up with more trouble out there?" I suggested. "Soon you might long for all the problems down here in comparison to the problems in the tin can."

He went silent for a while. The only sound was the clicking of the indicator as we turned onto Bruntsfield Place.

Eventually he spoke up. "You're right for sure," he admitted. "I'd soon long for home. But for a while I'd be somewhere else, experiencing somewhere different."

It fascinates me, the enduring pull of the escape. Even faced with a tin can containing no one you'd want to spend any length of time with, the possibility calls like a siren to the wandering spirit. Here on Earth, about 10,000 years of social, political, and economic effort have gone into our entrenched views. Although diverse in their substance, all of these views are bound by the human experience of Earth. It is perhaps not surprising that space would seem to offer the possibility of something genuinely novel—a society unlike any to be found on the home planet. The thrill of the frontier has an Elysian tint.

When Captain Kirk announces his mission to explore strange new worlds and seek out new civilizations, we are drawn into an idealized future of interspace travel where the petty exigencies of economics seem to have vanished, where battles with in-laws and tax collectors have disappeared, and all we have left to think about is exploration. The mere act of advancing into space is made to be a panacea and an act of liberation. But how realistic is this vision? If humans one day become extraterrestrials, will our off-world societies be free or tyrannical? What forms of government will be best?

Such questions lack the romance of escapist fantasy, but we should not pretend that there won't be politics in space. Although politics might seem far removed from aliens, the societies that humans build are very much part of the wider question of life in the universe and of how life—in this case ourselves—will adapt to new frontiers. When humans go into space, we will bring with us age-old questions about how to govern ourselves and achieve a good society. The collective problems of living in space and on other worlds are vast. To begin with, no matter where you go in space, the conditions are extreme and will therefore demand complicated remediation that no one person will be able to achieve on their own.

Consider that we know of no planet, certainly not in our own solar system, where a human can stand and breathe the atmosphere without considerable technological support. This single observation brings us at once to our point. To live elsewhere in space, we will need institutions that supply our most basic needs—institutions that are not necessary on Earth. The effort will be herculean, although, importantly, not impossible. On our nearest neighbor, the Moon, we could get oxygen from water, which has been found in the south polar region. There, deep craters remain permanently darkened, and water that otherwise would be vaporized by sunlight lingers in the soil. We could dig that dirt and warm it up to release the water. Then, cleared of dust, the water can be submitted to the process of electrolysis, in which an electrical current breaks it down into its constituent atoms to make hydrogen and oxygen. The hydrogen can be used for industrial processes, and the oxygen is ready for our lungs. Now we have some air to breathe.

You will have immediately noticed that this quite involved process puts a great many people between the oxygen source and a fresh breath of air. First, we need someone to go into the rocky lunar craters and dig the soil with the water in it. Robots will mitigate some of the danger, but they must still be superintended by a human operator, and spare parts must be provided for them. There will be a lot of activity to coordinate. Once we have the icy dirt, someone has to process it, itself a multistep endeavor involving cleaning, extraction, and filtration. Still others must oversee the water's transport to an electrolysis unit and its transformation into oxygen. Then we must have pipes that transport the oxygen to habitats and working spaces. The pipes and pumps will need servicing.

As much as we rely on water and electricity providers on Earth, the lunar oxygen utility will be still more essential and therefore more precious. Without water or power, life can soon become miserable, and for some, such as ill patients who depend on breathing machines, any downtime could be lethal. But most of us can last long enough for restoration of these vital services. The same would not be true of oxygen provision. Without it, the lunar inhabitants will instantly die. Thus does oxygen become a political problem in space. Whomever controls any of the technology and logistics standing between oxygen atoms and the human consumer will be powerful indeed. Each step in the process is an opening for a tyrant.

It is a grim thought, for who would want our cosmic future to be even more desperate than our earthly past? Throughout human history, control of resources has been a goal of authoritarians. Food, metals, water, land, fuel—all these and more have served to channel power into the hands of a dominating few. But never has anyone had the means to control the air we breathe. Thus, when societies have confronted appalling despotism, brave folks have been able to flee. They could build new homes or plot revolution. But when the air itself is controlled by a phalanx of functionaries, the capacity to resist is greatly reduced. Confront the authorities about the repression premised on their mastery of oxygen, and they can respond with an insincere apology and an offer to open the airlock so you can enjoy a second or two of freedom on the lunar surface.

Similar tyrannies could arise on other celestial bodies, even where conditions are less extreme. Mars, as we saw, has an atmosphere, though we can't breathe it because it is almost entirely carbon dioxide. Still, the task of gathering oxygen could escape centralized control. Instead of going through the trouble of harvesting oxygen from water, on Mars, atmospheric carbon dioxide can be directly broken up in chemical reactions to release oxygen. Perhaps everyone can own their own carbon dioxide–cracking machine, releasing them from fears of domination. But not so fast. The machines must still be fabricated, distributed, and serviced regularly. Mars's residents will still be uncomfortably at the mercy of the air-makers.

Food and water, too, will be levers of power in space. To grow a single stalk of wheat on the Moon is no simple matter. First, we must enclose a space with a greenhouse-like structure that will provide an atmosphere. As the Moon has no atmosphere to speak of and is, to all intents and purposes, exposed to the vacuum of space, this structure must be able to withstand pressurization so that we can inject into it a quantity of air sufficient for plants to grow. We must also regulate the greenhouse's temperature. The lunar surface soars to well over 100°C under the full glare of the sun; during the polar night, which returns every two weeks, the surface temperature plunges to below –150°C. Consigned to such extremes, our seeds would be dead before they could germinate.

The soil itself is not so bad. Made of volcanic basaltic rock, it contains plenty of nutrients. On Earth, volcanic terrains are some of the most fertile. But there is a snag: lunar rock is devoid of nitrogen, which plants need to live. In addition, the soil has been pummeled into an abraded dry material. These deficits can be ameliorated at considerable cost in fertilizer, perhaps with the help of a human-waste supplement. Then we must feed our seeds and the newly erupted shoots plenty of water. We already know how hard that is.

On Mars the task, again, is a little less formidable. The pervasive presence of ground ice simplifies the acquisition of water, and plants in a pressurized space should flourish off the atmosphere rich in carbon dioxide, which is the breath of life to plants. They can photosynthesize the carbon dioxide, turning its carbon atoms into sugars and new biomass to supply ever-hungry humans. Make no mistake, though; this is still a tricky task. Like the Moon, Mars suffers from high radiation. Ultraviolet light streaming in from the sun, unmitigated by Mars's nonexistent ozone layer, will burn a plant a thousand times faster than on Earth. Our greenhouse material must be glass, which naturally blocks harmful UV light, or else UV-resistant plastic. Of course, neither material is easy to make in space. All this is to say, providing a simple salad will demand considerable human effort across numerous disciplines.

Such effort could be mustered. And none of the relevant technology is beyond our capacities. It may not all exist now, but in theory we could develop all of the tools, chemicals, and materials required. But the greater challenge may lie in fostering a society capable of sustaining the interdependency without which we will lack even the basic predicates of life. There are so many pressure points in a lunar or Martian life-support system. So many opportunities for an individual or organization to hijack a population's capacity for subsistence.

Precisely because so much is at stake should any life-support system become compromised, an extraterrestrial society is likely to be marked by intense surveillance and a culture of vigorous command. Neither dissent nor even whimsy has much place where a single loose bolt could bring catastrophe. Earth has its hazards, of course, but they are as nothing in comparison. On our home planet, we can put up a sign to warn of falling rocks or rip tides and leave people to police themselves. In space, a wrongly pressurized habitat or a badly maintained airlock threatens instant large-scale death. Under such circumstances, it is easy for authorities to justify draconian control. Better to be safe than sorry—and woe betide the extraterrestrial settler who questions an order from those who know better, jeopardizing the lives of countless souls. The population will readily be coaxed into patrolling its own, as the environment itself will serve as a common enemy against which the whole of society is aligned. All must partake in this struggle if any are to survive. Dissent is an invitation to death.

Within such a society, individual agency will probably give way to acquiescence before authorities because the population will have little choice but to defer to those tasked with keeping them alive. Such circumstances run strongly counter to political liberalism, which regards autonomy as foundational to a good life and an open society—the sorts of ideas promoted by political philosophers such as John Locke and John Stuart Mill. It turns out that autonomy is contextually dependent, though. On Earth, where food and water are prodigiously available and there are no restrictions on the air we breathe, opportunities for independence are relatively easy to come by. Where these necessities must be acquired in complex collaboration with others, the sphere of one's will cannot help but be greatly diminished.

With liberty naturally restricted in extraterrestrial environments, the door to authoritarianism is open wide. Settlement managers will have many dependents to do their bidding, leading to a flourishing of tyranny. Who could prevent them extracting fealty, even servility, from their charges? A despot would not need even passport restrictions or physical barriers like walls and fences, because there is nowhere to go—no isolated forest or cave where the resistance might gather. To escape would require a spaceship, which is not the sort of thing that is easy to come by, especially since we can assume that the authorities will also control that sort of transit.

As we drove through downtown Edinburgh—a lively place, teeming with people and shopfronts—I engaged my driver again. "Don't you think that it would become quite suffocating to be with a very small group, let's say for your whole life, in a space station?" I asked. It should be clear by now that I am not a space-pessimist. I would go to Mars at the drop of a hat. But I think it is important to be clear-eyed. Did my driver have a rose-tinted view, or would he entertain the possibility that escape might turn into anything but?

In fact, my driver turned the tables on me. "Yes, yes," he replied. "But being dependent on people, needing others, can be a good thing. There'd be camaraderie, a common bond between everyone. It would be a real mission feel," he suggested.

I think this is quite right. Being in space would foster a strong sense of community and from there a different kind of freedom. This is the freedom to do together what no one can do alone. Through efforts to wield the resources of their settlement against the lethality of the space environment, settlers will experience this collective liberty born of common effort, common cause, and perhaps the government of all on behalf of the group.

Although the liberal philosophers who helped to create notions of freedom as the purview of the unfettered individual were rather skeptical of this collective kind of freedom, the ancients possessed such a vision. The ancient Greek city-state, the polis, was not defined by our contemporary notions of individualism, so focused on the exercise of personal freedom in spite of its potential dangers for the wider society. Rather, in the polis, the citizen achieved full potential by active involvement in politics. Without the polis, the isolated human was nothing. Such collectivism was necessary to flourishing, given the minimal population and the absence of the sort of infrastructure that we take for granted in our modern world. At its height, ancient Athens was home to just 140,000 people. Such a small group could not have run an empire if each person had not pulled their own weight on behalf of the group. Other long-ago empires, like the medieval Mongol Horde, managed comparable or greater feats through the subordination of the individual to the wider social scheme. By throwing oneself wholeheartedly into the social enterprise, the individual avoided relegation to a shriveled loneliness in which their own ambitions were inevitably stymied. Instead they achieved their potential within the collective.

Now I think you, reader, might agree that there is merit in this view. Even the most individualistic of us moderns benefits from social coordination. Without it, we couldn't fly away on holiday, buy the food we desire, and watch the films we enjoy. I need not belabor the entire network of collective effort needed to make a plane fly safely from one city to another. Today we hold notions very different from those of the ancients, but we are still part of the polis. It is just that the sheer size and extent of the collective makes it invisible; we often see in front of us only our own individual objectives, without realizing that we would have no chance of achieving them without the advantages wrought by group action.

The settlers on the Moon or Mars might build their own variety of the Athenian polis, and they will not be able to pretend otherwise. Huge efforts will be needed if the settlement project is to succeed, and the small size and high density of a population confined to a complex life-support structure will mean that everyone will work in proximity. It will be impossible to shirk or to look away from the management of public affairs. Perhaps the sheer difficulty of survival will nurture a view of liberty less committed to individual ambitions and more to those of a community on which each relies. If that is the end state of an extraterrestrial polis, then perhaps all is well and good. Perhaps the solar system will become a collection of Athens-like city states, bound together in an extraterrestrial Delian League.

And yet solidarity, too, has its discontents. My driver understood this. After a period of quiet, he spoke up again, but this time with more ambivalence. "I bet there would be a lot of peer pressure up there, being in a small group," he offered. "You'd have to fit in. And if you didn't, there'd be no getting away. But I suppose it would be nice to just accept it, you know, just be part of that and not have to worry about things."

There is indeed solace to be found in a sense of belonging through submission to the social order. But there is also peril here. When we abandon moral responsibility, instead allowing a higher authority to make choices for us, the results can be grim indeed. Leaders take advantage of public acquiescence to impose their notions of order. Hannah Arendt famously agonized over interviews with former Nazis to understand why they had submitted voluntarily, even eagerly, to an organization whose end was tyranny, and she found a haunting answer: almost all of her subjects despaired at the challenge of being responsible for themselves. To make decisions, implement them, and face the consequences of failure is taxing. By subordinating themselves to the will of others, and by allowing a totalizing ideology to provide easy answers, these Nazis were lightening their own loads. They were free—free of responsibility for making difficult life choices. Their failures were failures of the larger system, over which they had no control.

This powerful irony, whereby one experiences release by subordinating oneself to unassailable power, is at the root of so much human suffering. There is no reason to believe the root will die in space. On a space station or lunar colony, where one overlooked safety check could produce terrible losses, personal responsibility potentially is the stuff of nightmares. Some settlers will surely find it easier to abrogate responsibility, prophylactically absolving themselves while seeding the ground for despotism.

Perhaps, then, a space tyrant will not need to work hard to secure the conformity of the herd. Instead the sheer viciousness of the extraterrestrial environment will encourage many into voluntary servitude, as each accepts the freedom of escape from the terrible consequences of responsibility.

The beneficent dictator merely wishes to save us from a life that is nasty, brutish, and short. Why not partake of the liberty of submission?

It would be easy to lose heart at this point. If space is a place of tyranny, we needn't bother with dreams of settlement, much less actual attempts. So let me finish by saying that I do not see matters this way: I do not see tyranny as the inevitable result of extraterrestrial settlement. There is every possibility that the space frontier will birth new social formations more conducive to universal flourishing than anything humanity has created so far. Space is a tabula rasa, a perfect environment for what Mill called "experiments in living." Departure from Earth may well nurture new forms of art, music, science, and much else, alongside that new form of society in which each thrives. But we should not be blind to the dangers. We know that we are deeply fallible creatures, and it is clear that the conditions in space are suited to our worst instincts. To say that space is a natural home for despotism is to speak frankly, if not to predict the future. We should do our best to live with this truth by taking it seriously and doing all we can to develop governance that encourages the promise of extraterrestrial settlement. For while we may, in going to space, avoid some of the problems of Earth, we will not suddenly bring light to the darkness that lingers in the human mind. It will come along with us when we blast off.

Polar bears and lions benefit from environmental-protection schemes, but what about cyanobacteria, like this *Nostoc* colony? The individual microbes comprising the colony are just a few microns wide. Should we care about them?

14

Do Microbes Deserve our Protection?

A taxi ride from Bruntsfield to Fort Kinnaird.

✳

"You've cleaned your taxi," I remarked, as I slouched into the squeaky-clean black seat. A whiff of disinfectant hung in the air.

"Too right," my driver snapped. "Last night I had a lass in here who sicked up. She was on the bender, gets in, and then was sick about two minutes later. God I hate these Friday nights. Don't get me wrong. I don't mind people out enjoying themselves, but not when they can't hold it and I have to clean up the mess." She was almost turned right round in her seat, facing me, while she gesticulated toward the seat in frustration. As she spoke, her bushy brown hair bobbed up and down and her face set in a grimace. Her wide, shoulder-padded blue-and-white-striped jacket and middle-aged seriousness added to the sincerity of her irritation.

"It's probably a good excuse for a cleanup," I said. "I can't remember when I did the same in my car. I think never." I was slightly taken aback by my sudden introduction to the concerns of her world. She shook her head, and I thought perhaps she could do with a philosophical distraction. Only the day before I had been reading a scientific paper about whether we should kill microbes on Mars if we were to find them in a base on the planet.

"Would you have disinfected your taxi if it was full of alien microbes?" I asked. "Microbes from Mars?"

She didn't answer. She was looking sideways into her mirror at me and could tell that I was not joking, but rather staring back at her, waiting for an answer. "Are you serious?" she asked. "You mean like if I had some aliens in my taxi. Would I clean it out still?" she asked.

"Yes, yes, exactly," I replied. "If you found out that the sick woman had dropped a container of rare Martian microbes in your taxi. Would you have still cleaned it out?"

"Bleach them. I mean, just because they're Martian, who cares? I'd bleach them anyway."

"Even if they were different, I mean really different Martian microbes?" I couldn't really let go.

"You seem to think they would be interesting, but I'd disinfect them."

She sat silently and looked at me again in the mirror. She was quite fed up with the last twenty-four hours, and I was contributing as much to her exasperation as had last night's drunk passenger. My driver's attitude toward microbes was understandable. If I was her, I would have bleached my taxi as well. But when we are not cleaning our kitchens and taxis, there is another side to the story.

Turn up at an environmental rally. Or sit outside the United Nations during a climate change meeting. Look out for the signs: "Save the microbes!" "Justice for fungi!" "I'm with the slime molds!" Well, you won't see them. And you won't see representatives of the Royal Society for the Protection of Microbes, or the World Microbe Fund, or any other name you can dream up for an organization that cares about the defense of microbes. For most people, the idea of protecting microbes seems absurd. We kill them every day. I don't know how many microbes you kill when you bleach your kitchen surfaces, but it's likely many millions. Anyone demanding the preservation of some threatened bacteria would seem deranged, or at least lacking any sense of perspective.

Yet these humble creatures are at the core of our biosphere. Unseen and usually unrecognized, they are heroes of the living world. But we tend to think about them only when they make our lives a misery—when we can blame them for, say, food poisoning. Each year, in the United States alone, about 48 million people get ill from bad food, 128,000 of them end up in hospital, and about 3,000 die. So it's not surprising that producers of disinfectant take great enjoyment in announcing that their products kill "99.9 percent of all known germs." It is also not surprising that the word "germs" has become a pejorative term to describe microbes. The last thing I want to do is catch your germs.

In the seventeenth century, an inquisitive Dutch fabric maker named Antonie van Leeuwenhoek designed some tiny glass microscopes, which he used to look in exquisite detail at the quality of the cloth he was selling. He wanted to ensure that his material was the best available. But when he got a little bored with looking at fabrics, he turned his attention to pond water and even plaque scraped from his teeth. He was

astonished by what he found. What he saw with his microscopes were tiny animals—"animalcules," he called them.
These swarming, multiplying beasts captured the imagination of the age. Minuscule creatures, smaller than a hair's
breadth, opened the door to a Lilliputian world. For a while
they were seen as harmless and extraordinary, and their discovery was counted a triumph of scientific progress.

But then things turned sour. Robert Koch, Louis Pasteur,
and many others began to pry apart the world of microbes,
and soon a horrible secret was revealed. These flitting,
sprightly little life-forms were harbingers of the most terrible
diseases. For a century, the list of their misdemeanors grew:
Black Death, typhus, botulism, anthrax. Faced with overwhelming evidence, the microbial world could do nothing
but throw up its hands and plead guilty. "Your punishment,"
said the judge of humanity, "is that from this day forward you
will be known as germs and your presence among us will
bring you deserved opprobrium. An appeal is not permitted."
Well you could hardly blame people. Black Death wiped out
a third of Europe in the fourteenth century. That disease
alone would have consigned the microbial world to ever-
lasting infamy.

However, even as humanity was learning to decry microbes, some scientists were convinced that these creatures
had a Janus face: they could cause enormous strife, yet this
was not their sole lot in life. One of those scientists was Sergei
Winogradsky. Born in Kiev in 1856, Winogradsky was a brilliant, multitalented individual. After dropping his studies at
the Imperial Conservatoire of Music in Saint Petersburg, he

turned to botany and from there the universe of microbes. Winogradsky was one of the first to recognize bacteria's immense importance in the environment. In particular, he discovered that certain bacteria could derive energy from the element sulfur. This demonstrated that bacteria were not just passive riders, eking out a living in the microscopic unseen edges of our world. Rather, these creatures were a living part of the planet, shifting and churning the elements on which we all depend.

Consider one element your body needs in order to survive: nitrogen. The atmosphere is about 78 percent nitrogen, so there's plenty available. But those atoms of nitrogen are imprisoned, locked up in gaseous form. A molecule of nitrogen gas contains two nitrogen atoms, bound together so tightly that they cannot be separated mechanically. This is where our microscopic friends take center stage, for they can pull apart the atoms and rearrange them. Snapping apart the nitrogen gas, they attach some hydrogen or oxygen atoms to the resulting free nitrogen atoms, making ammonium and nitrate, forms of nitrogen that are much more easily consumed than is nitrogen gas; both ammonium and nitrate readily dissolve in water, and they like to take part in all sorts of further chemical reactions. This process is called "fixing" nitrogen. Each microbe is a tiny nitrogen-fixing factory, and, working together, they achieve amazing feats. A typical microbe is about a thousandth of a millimeter long—a micron—but there are vast numbers of them. Collectively they extract a staggering 140 million tons of nitrogen gas from the atmosphere each year and turn it into fixed forms of nitrogen that feed the biosphere. So while it is true that microbes can menace our health, it is also true that, without them, we'd all be dead.

Work done by Winogradsky and others demonstrated the extraordinary reach and power of the microbial world. Beyond fixing nitrogen, microbes do many other essential tasks. Fungi decompose dead plants and animals, returning their bodies to the biosphere to be used by the next generation of living things. Alongside sulfur, bacteria cycle carbon, iron, and almost every other crucial element of life, round and round in a planetary biological Ferris wheel that keeps our biosphere from sputtering like a dying engine. Closer to home, we have microbes to thank for fermenting sugars into wine and beer, pickling vegetables, and working their magic on milk so that we can have yogurt and certain cheeses. Today we also use microbes to make the drugs we need to fight the very diseases caused by their kith and kin. And let us not forget the vital role that microbes play in our bodies. They help digest our food, breaking down meat and vegetables, activities that make them essential to our health. About half the cells in your body are microbial; you just can't see them because they are smaller than human cells. It always comes as a bit of a shock to people to learn that they are only half human, at least from a cellular standpoint.

It is difficult to countenance this benign view of microbes given the destruction they let loose on our world. It's like trying to forgive a serial killer. But the fact one has to come to terms with, however difficult, is that no number of microbially caused human deaths can alter the immensity of their roles in keeping the biosphere ticking.

So where are all those T-shirts yelling "Save the microbes"? Well that's a fascinating question, and I'm not sure anyone has a complete answer. It might be that many people, inclined to side with the prosecution, think that microbes just don't deserve protection. Lumped in with that, there are just so many of them; perhaps they don't need our help. There are thought to be fewer than 4,000 tigers in the world. How many microbes? Researchers argue about it, but recent estimates, based on counts in the oceans, soils, and all the other microbe habitats, suggest that there may be about one thousand billion billion billion microbes on Earth. Endangered they are not, so few see any need to care.

Microbes also do not look very attractive, which has a great impact on human sympathies. How many T-shirts do you see with "Save the small endangered spiders!" or "Save the endangered gut parasites!" on them? Microbes aren't the only ones to get the cold shoulder from environmentalists and, if you'll forgive me for saying, most people get worked up about seals, pandas, and other creatures with a cute face. As environmental ethicist Ernest Partridge once pointed out, anything with the "aw gawsh factor" will enjoy the most attention from concerned humans.

Perhaps the best explanation for our lack of compassion, despite all that microbes do for us, is just that they are difficult to see. Out of sight, out of mind is a literal reality for them. Imagine if polar bears were a thousandth of a millimeter long and microbes the size of a dog? The poor microbes would probably still be quite ugly—bags of water bobbing around in a lake, a few of them with strange whip-like appendages thrashing about and making a terrible gurgling noise as they swim around nibbling bits of food. No doubt some people would feed them like ducks. But at least you would see them, and every time you drained a lake, their fate would be horribly real. At the same time, you'd be much less likely to campaign for the preservation of polar bears in this alternative world, where their microscopic millions wriggle invisibly in spoonfuls of soil. Of course, polar bears the size of microbes are not physically plausible, but that isn't the point. What we should take from this thought experiment is that size is part of what makes animals salient to us. The tininess of microbes probably has a lot to do with our hesitation to take their preservation seriously.

And there is the power of consumer society. When every cleaning product advertises the maximal killing of microbes, we learn that they are not worth saving. Bleaching them without remorse reinforces the view that microbes are our enemies, whereas much of the time they are our partners.

Are the microbes a lost cause? Not so fast. Dotted around the shoreline of Shark Bay, Western Australia, are strange domed structures. Brown, black, and blue, and up to a meter across, these wartlike protrusions, known as stromatolites, sit on the sand as the tide washes back and forth over them. Stromatolites look like rocks, but in fact each is made of layer upon layer of bacteria—cyanobacteria to be precise—sandwiched between grainy layers of sand. And, as living things, stromatolites grow. While the sand sinks downward, the photosynthetic cyanobacteria move upward to capture the sunlight, extending the mounds. The stromatolites are the jewel in the crown of the Shark Bay World Heritage Area, marketed to the fascinated public as "living fossils." Actual fossils of them have been found in rocks over 3.5 billion years old. To observe stromatolites is to take a journey to the dawn of life on Earth, when there was nothing more on our planet than microbes, and animals would not appear for another 3 billion years. I applaud the Australian people, for here is a real instance of microbial conservation.

About twenty years ago, science fiction writer Joseph Patrouch wrote an amusing story about a future dystopia where microbial rights are fully recognized. Deodorants are banned, and you can't clean your house. You aren't allowed to wash your hair. This is satire, of course, a glimpse into the absurdity of protecting microbes as though they were tigers. Yet the stromatolites of Shark Bay are indeed protected microbial communities. Large, visible to the naked eye, possessing their own form of beauty, and wowing us with their persistence, the cyanobacteria in Shark Bay have shown themselves worthy of human care, an honor denied to nearly all other microbes.

How do we reconcile the protection of the Shark Bay stromatolites with the hygienic killing of microbes? Some of our choices are perhaps not for the best: millennia of humans did without deodorants, and while their invention has convinced us of the value of smelling a certain way, we probably could do without them. But cleaning our homes or filtering our water is not merely a matter of avoiding odors deemed foul. Our health and lifespan have considerably improved with developments in sanitation. So perhaps we should take the view that we ought to protect microbes when we can, but we are not obliged to all of the time. Similarly, many of us would oppose the gratuitous destruction of trees, but we probably don't oppose cutting down some trees to get timber or make paper.

Perhaps we would err more on the side of protection, though, if we paid more attention to the useful things microbes do. We could return to the work of pioneers like Winogradsky and realize the vital role of microbes in cycling elements, decomposing waste, and generally ensuring the health of ecosystems. Microbes are the first link in the food chain: capturing sunlight, fixing nitrogen, and collecting all manner of other elements used by everything else in nature, microbes are truly foundational to all life and could equally be the foundation of an environmentalist ethic. We tend to think about the fish and frogs disturbed by water pollution, but, more to the point, pollution kills many of the plankton and other microbes that inhabit waters, and it is their loss that ultimately harms the larger life-forms we can see. We don't need to save microbes for their own sake, but protecting them is good for all other life on the planet. Microbes are the unseen girders in the biosphere, the structural supports tucked away in walls and ceilings that nonetheless hold the building up.

We could foster an environmentalist appreciation of microbes without descending into absurdities about banning bleach. For instance, instead of draining any and every pond to build housing estates, we might take a more nuanced approach that accounts for the local microorganisms. Yes, some ponds really are nothing special. But others contain rare and important microbes. If we took more seriously the crucial functions of microbes in maintaining our ecosystems, we might adjudicate more effectively among bodies of water and site our housing estate accordingly. If we can be awed by the stromatolites of Shark Bay—and we should be—we can learn to esteem an unassuming but essential local lake as well. Perhaps a T-shirt with "Save the microbes!" emblazoned across it is not such a silly idea after all.

What about the really nasty microbes, though? The big killers. Is it acceptable to hound, say, smallpox out of existence? For most of history since the third century BCE, smallpox was, well, a pox on humanity. This viral disease has been found in Egyptian mummies and killed nearly 300 million people in the twentieth century alone. Smallpox caused blindness in about a third of sufferers and spread boils across the skin. Yet today, we do not have to worry about smallpox. It has in fact become difficult to understand the scourge this disease once was—a terrible daily reality across the ages. We have the science of vaccines and the efforts of the World Health Organization to thank. Beginning in the 1950s, the WHO led a heroic global initiative to eradicate smallpox by inoculating the human population. It was the first planetary-scale war against a single microorganism, and it worked marvelously. The last naturally occurring case of smallpox came in Somalia in 1977. The infected person, a hospital cook named Ali Maow Maalin, survived, recovered, and went on to become a vaccination campaigner.

I completely understood my driver's intention to bleach anything that could be dangerous, but what about driving things to extinction? "Do you think," I asked her, "if we could corner the last of a really nasty disease-causing microbe, like smallpox, we should finish it off?" There was a brief silence as she shook her head in consternation. "Kill it off," she exclaimed. "Why would you go to all the effort to round it up and then let it loose again?"

I'm sure that many would agree. But consider that a deliberate and concerted global project to snuff out the last tigers and elephants—both of which can be dangerous to humans—would be considered insanity. Who are we to decide among these creatures that share a planet we do not own? What arrogance we possess, presiding over the extinction of creatures, choosing which will live and die. Some species that we might knowingly eradicate have inhabited Earth for hundreds of times longer than our own. Why doesn't smallpox have a right to continue to exist alongside tigers and elephants? I think my driver's reaction was understandable, but the correctness of it isn't necessarily obvious.

As it happens, smallpox was not made extinct. There are vials of it at the Centers for Disease Control and Prevention in the United States and at Vector, Russia's State Research Center of Virology and Biotechnology. Although the date for the final destruction of smallpox was set for December 30, 1993, it seems we are too frightened to let it go. Perhaps, the thinking goes, it will reappear somewhere, and then we will want a stock of the virus around so that we can research it and combat potential outbreaks. Smallpox got a stay of execution precisely because it is so awful. But within smallpox lies the central conundrum: When does a creature become so vile that its deliberate destruction is merited? At what point is any intentional extinction acceptable? Microbes have never made ethics easy.

The matter becomes even less clear when we take a speculative ride to other worlds. Arguments about smallpox and bleach take on a new dimension when we come upon life elsewhere. If I suggested bleaching all the microbes on Mars to make way for our stations, I think you'd be appalled. But stop for a moment. Why? You bleach your house; why do the microbes of Mars get special treatment? I suspect you would respond with arguments to the effect that these microbes were just doing their thing, living happily on Mars. Who are we to turn up and kill them all?

Embedded in this view is a respect for life, a respect for the microbes of Mars that puts their existence above our own interests, above our instrumental uses, as ethicists would say. This sense of respect is difficult to pin down, and ethicists have enormous challenges defining it without falling prey to sentimentalism. But I think it does capture something fundamental about how we think of living things: a conviction that other life, going about its existence however blindly, has a right to carry on. Maybe this is an innate humility shining through. To destroy a whole biosphere on Mars, even if only microbial, speaks badly of humanity, reflecting a cruelty we don't want to see in ourselves.

Perhaps the idea of destroying Martian microbes stirs within you the same feeling you might have if you saw a nonchalant teenager on a tour of Shark Bay stride out onto the beach and start to jump up and down on those stromatolites, destroying them one by one. You can't tell me that your outrage would be caused by your natural affinity for cyanobacteria. Only the weekend before, you probably washed a bunch of them off your car and consigned them to oblivion. No, your outrage likely stems from the teen's behavior, the disrespect, the gratuitousness of the destruction. Feeling this way doesn't entail a commitment to protect all microbial life; there are other interests we might justifiably rank above those of the microbes. But the stromatolites, at least, deserve to be left alone because we have a sense there is some inherent value in living things. Something truly worthless cannot be gratuitously abused.

We haven't found alien life yet, but merely thinking about microbes elsewhere can motivate us to contemplate our relationships with the natural world. Things we take for granted, like bacteria, can suddenly take on an importance that we don't consider in our day-to-day lives. When we think of them crawling through the sands of Mars and wonder how we would treat them, we gain a fresh perspective that can help make sense of our actions toward them on Earth.

I agree with my taxi driver when it comes to cleaning my house and car and washing my hair. I laud the extraordinary advances in medicine that allow us to kill the bringers of microbial disease. I stand shoulder to shoulder with scientists who work to overcome our antibiotic-resistance crisis, because we need to find new ways of checking microbes that have evolved to sidestep our protective medications and vaccines. But I also love the microbial world. Among the many types of microbes on Earth, it is but a minority that happen to cause trouble to people. This is no reason to hate them all. I love tigers, although I know that some have attacked humans.

Microbes' persistence on Earth for over 3 billion years, their irreplaceable contributions to the biosphere, their thankless labor in creating a world fit for our existence—these are good reasons to respect microbes. Indeed, I admit to a certain reverence for them. There is plenty of room on Earth, to say nothing of Mars, for the donning of "Save the microbes!" T-shirts. And yes, there must be justice for fungi too.

Life emerged in Earth's early history, perhaps in a hydrothermal vent similar to this one—the so-called Candelabra, which spews hot fluids into the sea at a depth of 3,300 meters.

15

How Did Life Begin?

A taxi ride from Oxford station to Corpus Christi College.

✳

I didn't have far to go, but it was raining, so I jumped into the back of a taxi at Oxford railway station. I was on my way to Corpus Christi College, where I did my doctorate, for a celebration of the 500th anniversary of the institution.

"Is Oxford home?" my driver asked. He was probably in his sixties, and I immediately sensed he was an old hand at the job. His head flicked from side to side as he carefully navigated his way past some parked taxis along the edges of the station and out to the main road.

"It used to be, but not now," I replied. "But it always feels a sort of home, maybe even melancholy to be back."

There's something unsettling about traveling streets trod in your youth—here you walked late at night after a party, here your young mind worried itself with one source of angst or another. These streets are occupied by your ghosts. I lit up another conversation with my driver about how I spent some of my years here—just three of them, less than 1 percent of the college's existence. But even 500 years constitute an insignificant blip in the time since life itself emerged on Earth. Four billion years have elapsed since then; Corpus Christi has existed for merely 0.0000125 percent of that time. On that scale of things, I am a more significant presence in the history of the college than is the college in the history of life on Earth. (And I am not, after all, especially important in the grand scheme of Corpus Christi.)

"It's easy to grasp that my own time here was a tiny part of the college's history," I told my driver, "but if you'll forgive me for a strange thought, it's even more humbling to think about the college as a minuscule part of the time since the birth of the Earth billions of years ago. We are all quite ephemeral compared to that."

"It's difficult to comprehend times so long. It's just not what we think about normally," he replied, as I nodded in agreement. The human mind struggles with the meaning of centuries, let alone billions of years. That is just a fuzz of time; we find it hard to really grasp the difference between a million and a billion. These are the sorts of spans that were crossed while life emerged. Was this inevitable, given the time available? Is Corpus Christi an intellectual home to flukes of chemistry, or is it certain that living things of some sort would have emerged from the bubbling, simmering ponds of primordial Earth—life that would one day possess intelligence? As we discussed my interest in the origins of life, it was this question of inevitability that occurred to my driver.

"Time is so long," he said, "almost anything could have happened. How did this all begin and was it for sure?" This is, or should be, a question fascinating to all, though it is answerable by none. Still, we don't often ask it so directly. Another driver in London had asked me whether alien taxi drivers were scattered across the universe, and underlying that question is the further enigma of the inevitability of life.

But this time my driver homed in on the fundamental issue of how it all began. Once the blistering-hot Earth solidified from the molten rock of our early solar system, was it preordained—not necessarily by God but by physical conditions—that the planet would become infested with life?

Any possible answer to this question is inextricably connected to the existence of life elsewhere. If life is inevitable given appropriate conditions, then Earth is not likely a freak show in a dead universe, because it is just inconceivable that, among the billions upon billions of planets in the universe, none other is like Earth. Indeed, we should not presume that earthlike conditions are the only life-supporting conditions, and it is still harder to believe that in an Earth-containing universe, no other planets harbor any variety of such conditions. My taxi driver's question—how did this all begin and was it for sure?—may sound rather esoteric. It may seem like the sort of conundrum that one contemplates in one's adolescence before submitting to cynicism or responsibility in adulthood. Yet that question is of the utmost significance, and it has inspired countless generations of scientists.

I am not, in this chapter, going to let you in on the secret of whether life is inevitable. But if no one can honestly claim to know one way or another, I can at least try to explain what we do know. Serious research has produced tantalizing ideas. We have been able to rule out some theories—an important part of the scientific method, you'll recall—and continue to test others and follow where they lead.

Here is one approach to thinking through the origins of life. Dissect the simplest bacterium on the planet and you will find that it has some basic parts shared by all life on Earth. In other words, there is a sort of fundamental plan of life, a bit like how every car has certain features in common. Amid a profusion of forms and colors, every car has an engine, doors, wheels. And so we find at the heart of life a chassis that is invariant across the biosphere. It is quite reasonable to ask where these bits came from, for they are evidently the building blocks of everything that came after. And if we understand how these rudiments of life came to be, we can more effectively search the cosmos for other places whose conditions are conducive to their emergence.

One of these essential features of life is containment. Every organism on Earth has an interior, bound within a surface that separates it from the surrounding environment. And within this interior space, in many cases, are lots more objects with their own interiors, themselves bounded and separated from surrounding tissues. This is a clever solution to one of the problems encountered on a planet mostly covered in oceans: in water, things tend to disperse. If you put a small amount of dishwashing liquid into a tub of water, the soap dissipates and mixes until its color is barely perceptible. Similarly, try to concentrate some of life's essential molecules in an ocean, river, or lake, and they too will become diluted. The only exceptions are viruses, like coronaviruses, and prions, the errant proteins that cause mad cow disease and other illnesses. But these can't replicate on their own. They are desiccated particles if you like, and they depend on the liquid-filled interior of the rest of life's compartments to become active and replicate.

Indispensable to life, then, is a bag to put all its stuff in and prevent diffusion. Such enclosures exist at all scales of life, but at bottom—the one that started it all—is the cell membrane. Every cell on Earth is contained by such a capsule, and

. although these capsules vary across the zoo of life, they share a common feature: they are constructed from a particular type of molecule. These molecules, called phospholipids, have a head and two tails. The head is hydrophilic—it "loves" water and will happily make contact with it. The two tails, however, are hydrophobic; they are repulsed by water. If you add some of these molecules to water, they spontaneously do something remarkable. They form a sphere with the hydrophilic heads on the outside, exposed to the water, and the hydrophobic tails pointing inward, shielded from the water by the surrounding heads. Meanwhile, other phospholipids arrange themselves within this sphere, facing the opposite direction, so that tails face tails and the interior heads enclose a hole containing water, and potentially much else besides. It is an astonishing transformation but not a miracle. Rather, the formation of the phospholipid bag is a simple and ineluctable consequence of the character of the phospholipids themselves, with one end that likes water and another that hates it. It turns out that one of the best ways for such structures to coalesce together into a stable state is to align into sheets and then collapse into spherical forms. This structure, called a vesicle, is both beautiful and vital. It encloses all the odds and ends needed for life to exist. When a vesicle encloses a cell, we call it a cell membrane.

An obvious question is where these remarkable contradictory molecules came from. They seem rather specialized, tuned for the very particular needs of making life's membranes. In fact cell membranes are finely tuned; in the course of evolution, they have become more suitable for life. But they have their origins in something simpler—the primordial material from which our solar system was constructed.

Get yourself an ancient meteorite that includes some carbon-containing molecules. A good example would be the Murchison meteorite, which landed in 1969 in Murchison, Australia. This meteorite is a relic of the fabrication of the solar system: more than 4 billion years old, the meteorite dates back to the very beginning of the story of our solar system and the life it would spawn. The rock is black and almost soft to the touch, its color caused by its high content of carbon-rich organic compounds, not unlike soot. Its appearance might convince you that it had even been burned at some point. Now gently crush the rock in water and release some of the molecules within. Among them are molecules with long chains of carbon atoms stuck together. Extract these so-called carboxylic acids from the mix and add them to water and before your very eyes they will aggregate to form vesicles that pulsate and float beneath your microscope. These molecules lack the complexity of today's cell membranes, shaped by more than 3 billion years of evolution. But carboxylic acids, brewed in the nursery of our solar system's clouds and gases, are the simplest bags of life.

Exactly how carboxylic acids formed remains something of a mystery, but we do know that the universe is full of carbon chemistry. Manufactured in the fusion reactions of stars, including the bizarre "carbon stars" that periodically convulse and throw off shells of gas containing a cornucopia of carbon-containing matter, carbon atoms are distributed liberally through the universe. Carbon monoxide, one of the simplest carbon molecules, can be found in abundance throughout interstellar space. A menagerie of other more complicated molecules—right up to the strange soccer ball–

shaped "buckyballs," or fullerenes, which contain a prodigious sixty carbon atoms or more—drift through the void.

Here and there in the universe, this carbon came into contact with other elements in contexts that spurred chemical
reactions. The nebula from which Earth and all the objects
of our solar system emerged was one such meeting place: an
interplanetary-scale chemical factory with temperature and
pressure gradients—and some radiation thrown in for good
measure—that resulted in a vast number of chemical experiments. Icy grains provided surfaces on which carbon chemistry could multiply the diversity of molecules on offer, potentially including the carboxylic acids that can form those
primitive membranes.

If all of this molecular chicanery seems complicated, well,
it actually isn't. It only sounds that way. The creation of the
first ingredients for life was not so difficult; the chemical
compounds and energy sources that together produce carboxylic acids are easy to find all over the universe. No one
need direct this process. Given appropriate matter and energy
and sufficient time, a bag suited to containing the molecules
of life can readily emerge.

But life is more than a bag. You need a few other things to
get a replicating cell. Particularly handy are some molecules
that can drive chemical reactions, speed them up, and produce a variety of other molecules that are rare or nonexistent
in the natural environment but which also are crucial for life.
These catalyzing molecules are the enzymes. They bind different sorts of molecules to each other and spew out the resulting products. Almost all enzymes, in every living thing
we know, are made from proteins, which are themselves

nothing more than long chains of amino acids strung together like beads on a string. The string folds together, transforming into a small three-dimensional molecule—a protein—ready to perform one or another kind of work.

Amino acids are simple molecules with an unpretentious structure made up of a central carbon atom onto which another chemical group (a small collection of atoms) is added. This molecular appendage comes in different structures, or flavors if you like, depending on what is attached. Different chemical groups do different things: some like water, some hate it; some have a positive electrical charge, others a negative charge; some are small, and some are bulky. The interaction of all these different flavors of molecules causes the long chain of amino acids to fold up in particular ways. Some strings of them will form a support structure, rather like a scaffold, which could be useful for building things like nails and hair. Other strings will take part in the all-important reactions that cells perform. Remarkably, all this can be done with just twenty amino acids. But as a protein can be many hundreds of amino acids long, you can see that if each position in that chain can have one of twenty different amino acids, the total number of protein combinations is enormous, much greater than the number of different kinds of molecules life could need to make a cell.

Let's take another look at the Murchison meteorite. When we did our extraction to get at the membrane molecules that make the cellular bag, there was something else astonishing that we would have noticed: the meteorite is full of amino acids. More than seventy different kinds of them, in fact. Amino acids, the building blocks of proteins, were synthesized in the chemical factory of the early solar system. They eventually were incorporated into the rocks and other material from which the planets were constructed. And some of the rocks kept hurtling around the cosmos, eventually arriving on Earth over 4 billion years later to be collected by scientists, who have glimpsed in them the simplest of life's ingredients—products of interplanetary chemistry.

The meteorites contain many more amino acids than the twenty needed by life. If indeed life got its first molecules from this interplanetary storehouse, why was it picky? The reason is that sometimes you don't need every material available in order to do a good-enough job. When architects design a house, they don't do their best to use every available variety of brick and roofing tile in existence. They pick a few to get the task done—often the fewest possible. This is the most efficient approach, and it prevents incompatibilities among building materials. In the same way, the fact that nature contains many more amino acids than those found in life is of no real consequence. Evolution is not a process of maximizing in every dimension; as long as a cell could meet its needs and reproduce, it didn't have anything to gain by incorporating more amino acids. It is understandable that from the rich chemical larder, life required only a selection of molecules.

Our extraterrestrial messenger has other surprises, too: nucleobases. Central to everything your cells, and even the most primitive life-forms, do is a code to store information; nucleobases make this possible. Most living things use the familiar DNA (deoxyribonucleic acid) for this purpose, or else its sister molecule RNA (ribonucleic acid). Like the proteins, these are made from long strings of molecules—the nucleobases. Unlike the twenty amino acids in proteins, only four nucleobases are needed in DNA. Scattered along the chain, the sequence of the four nucleobases is the code within which the instructions to build everything from eyes to tails is encrypted. The apparatus of the cell decodes the encrypted information, revealing the vaunted "blueprint" from which a whole creature is constructed.

Nucleobases were produced in chemical reactions involving cyanide compounds and other miscellanea of the early solar system. That's how they got into meteorites and other material hurtling through space. And, like the amino acids, there are more nucleobases in these interplanetary objects than are found in the living cell. Through the evolutionary process, their number was pared down until life on Earth was left with the essential requirements.

There is something extraordinary about all this. Peer through all the complexity of living things to the superstructure beneath it, the girders, bricks, and mortar that hold it all up, and you find that all the simplest parts of the major molecules of life are to be found in meteorites. These meteorites were there at the dawn of the solar system, no doubt raining onto the surface of our young planet, collecting in pools, washing up on beaches and in rivers. Scientists, enthralled by this possibility, have attempted to recreate in the laboratory the reactions that unlocked life's fundamental materials from their rocky vaults. Sure enough, when you irradiate the surfaces of mineral grains harboring simple molecules such as alcohols and cyanides, amino acids and other molecules important to life spring forth.

That is not all. Let's not forget that Earth itself is a space rock; the chemistry occurring on the land, in the oceans, and throughout the atmosphere of our newborn planet could have produced, and likely did produce, the stuff of life at the same time it was raining down from space. You can't easily escape carbon chemistry. From the depths of space and from within its own landscapes, Earth was infused with the simplest of life's needs, so that the scaffolding of biology accumulated on the surface of the planet from multiple sources.

All of this is deeply interesting, and it provides a plausible explanation for the origins of life's basics. But that still doesn't really answer our question—was life inevitable, given the right conditions? For why wouldn't these compounds just wash back and forth inertly in a tide, filling cracks and crevasses without so much as a cell in sight? This is where our research to date comes up short. There are a lot of clues and possibilities, but there is no consensus as to what threshold our life-sustaining compounds had to cross so that there actually was life.

Scientists have their conjectures, of course—hypotheses about what caused the crucial life-giving reactions to occur. These hypotheses are often related to particular places on Earth, locations where matter might have met energy in just the right way. Some scientists favor hydrothermal vents, fissures in the seafloor where hot fluids emanate from Earth's crust and leave towering mounds of minerals at the bottom of the oceans. Within these edifices, chemical reactions might have produced building blocks of life. And, more to the point, these could have been the birthplaces of the first metabolic reactions, which synthesized the molecules from which the great machine of biological energy production emerged.

Other scientists favor the beach. At every tidal cycle, waves crashing against the rocks would have delivered some of those essential amino acids from the ocean to a surface on which they could gather. As the tide receded, the dried molecules would have stuck together, the evaporating drops of water forcing the molecules into proximity and then union, each tidal cycle adding to a growing chain until eventually the earliest living molecules would have leapt forth from their rocky confines.

Still others eschew the rocks and look instead to the sky. Bubbles bursting at the ocean surface would contain on and within them the smallest molecules of life. As they drifted through the atmosphere and were exposed to the sun's ultraviolet radiation, reactions could occur, enabling mutations and kicking off the evolutionary process. Those newly complicated molecules then rained back down into the ocean to repeat the cycle, from which comes the stuff of life.

All of these ideas have some merit, and none are mutually exclusive. It is possible that at each of these points—deep-sea vents, beaches, the ocean surface—contributions were made to the early stock of chemicals from which life emerged. Perhaps the whole of early Earth was a gargantuan life-giving reactor.

At some point, no matter what your pet theory is, these molecules must have collected, or they would have suffered the fate of most things tossed into the ocean: dilution. Then, the early membranous molecules must have enclosed a molecule that was capable of replicating. Over time, the replicators would have taken control of the membranous bag, adding to it a new complexity of molecular form. From these simple beginnings, errors and variations would have produced a smorgasbord of cellular types, eventually leading to Earth's first cells.

What about this step, this lurch from a soup of chemicals to a replicating cell—was it inevitable? We just don't know. It could have been easy. Imagine an Earth covered in a slick of organic compounds raining down in meteorites and ushering forth from its own surface. Within the admixture of membranous molecules, a billion experiments occurred every day. Only one needed to make a simple cell that could replicate, becoming the unit that could evolve. Perhaps the emergence of life took less than a day.

Within these questions are a million other mysteries. Which came first, proteins or DNA and RNA? If proteins came first, how could cells have had the information needed to make them in the absence of an encoded blueprint? Alright, perhaps the code came first. But if the genetic code, that first flapping piece of RNA or DNA, was the progenitor of life, what use could it be, just a string of cryptic chemicals that coded for nothing in particular? Perhaps this early code was itself a chemical reactor, both code and catalyst in a single chimeric form. In that case, the catalytic proteins might have joined the party later, giving extra complexity and possibility to life's structure.

These mysteries could be interpreted in two ways. On the one hand, they could be telling us that the structure of early life was diverse. Protein first or nucleic acids first or both at once, operating independently. Maybe, with billions of experiments going on, it did not really matter. Every molecular permutation that could have emerged did, over and over, until somewhere in this planetary broth certain chemical combinations produced cells. Maybe early Earth was awash with a global competition among these primitive living things, any one of which could have been the first ancestor of all life that later called Earth home.

On the other hand, what we know, or suspect we know, about the early Earth is also compatible with the hypothesis that this stew of molecules was lifeless until one very special circumstance arose. On this view, proteins, DNA and RNA, membranes, and other components of life engaged in a constant dance as they were dragged around by the energetic forces around them until at last all aligned to produce the first cell. If this is the case, then Earth was still enveloped in a primordial organic goo, but the destiny of most of it was to remain moribund. There was no competition among forms of life, no hothouse of evolutionary experimentation. Instead, somewhere a random triumph occurred. Just the right components of life found themselves, for no reason at all, within a membrane. They did their thing, whatever that was, and bulged outward until, for the first time in Earth's history, a membrane split in two, so that each member of the new pair contained its own tiny, identical collection of replicating molecules. And from there, again. And again. And again. Now sixteen of these cells lived on Earth, and every few minutes they divided. And again, again, again. Now 128. And again, again, again. Now over a thousand cells. Within a day, a world had been conquered. Earth had become life; a biosphere was born.

Perhaps the universe is full of oceans lapping up against coastlines, and hydrothermal vents and bubbles daily circulate the building blocks of life from amino acids to membranes without so much as a cell in sight. Maybe none ever arise. The potential for life is there, for there is energy aplenty alongside the right chemical compounds. But, beyond Earth, the gulf between the ingredients of life and the reality of life itself is always as wide as that between a child's discarded building blocks and the cathedral they could become.

By observing other worlds and seeking life, and by continuing experiments in the laboratory, we may eventually fathom whether we are lucky or common, whether life is destined to flower from the molecular simplicity that pervades worlds like ours, or whether we are the children of a most extraordinary moment. Although the prospect of an intellectual and cultural exchange with intelligent aliens should excite us, there are also more basic scientific reasons to look for life elsewhere. We could find some extraordinary clues as to how our own world became what it is.

In our short skip from the train station to Corpus Christi, just a few minutes, I had no time to explain to my driver this long and uncertain history. When faced with the mystery of life's inevitably, I had to conclude in defeat. "I can't answer your question," I told him. "But no one can yet answer your question. That's why it's such a fascinating one. We just don't yet know whether life is rare or mundane." As we pulled up at the college, he smiled and shook his head, "Yes, yes, such simple things we just don't know." I nodded in agreement and thanked him. In that final comment he captured a truth as eloquent as it is uncomplicated. We have answers to lots of daunting questions. We can explain in detail many intricacies of our bodies, our environment, and our universe at large. But the really basic answers remain elusive. We can trace the evolution of species for thousands of generations, but we cannot say why there is life on Earth in the first place. I handed over my fare and let myself out into the street.

A beautiful sky can take your breath away, even as it contains the stuff you literally breathe. The oxygen in Earth's atmosphere is the invisible fuel that powers us and much of the biosphere, which is why scientists are looking for the gas on other planets.

16

Why Do We Need Oxygen to Breathe?

A taxi ride from Her Majesty's Prison Edinburgh to Bruntsfield, after delivering a class for inmates.

✳

It was a chilly morning, the air almost syrup-thick with humidity. One of those days you actually become aware that you live in an atmosphere.

"Cold day out there," my driver coughed as we drove out of the prison gate. I had just been working with some of the prisoners on their designs for a Moon base. The course is part of the Life Beyond project, an educational initiative that teaches inmates scientific concepts through the lens of space exploration.

"You can almost eat the air," I replied, perhaps bizarrely. "It's that cold."

"It's a funny old thing," he said, referring to the air. "We take it for granted that it's there." Then he asked me what I did for a living. I sensed that he was an inquisitive sort. Sometimes that quality jumps out at you in a taxi driver. You get into a cab, sit down, and as soon as you say something that veers off a mundane greeting, the driver sees a chance for a chat. This one sat up as he spoke, big bushy black eyebrows reflecting in the mirror while his arms sort of yanked on the steering wheel. He wore a yellow coat with a high collar that wrapped around his graying black hair. I explained my work to him.

"So you're a scientist. Tell me about air then. I mean how did it get there and why can we breathe it?" he asked.

It was a strange question to be asked in a taxi. In fact, a slightly surreal question under most circumstances. Having said that, the history of Earth's atmosphere is a fascinating one. I teach my astrobiology students about it each year, and after you've done that for a while, you can forget that it's not necessarily something many people give even a thought to—how all that oxygen collected in the atmosphere in the first place, free for us to breathe. But some taxi drivers wonder.

Stand motionless on a cold winter's morning looking out over a field, crisp frost in the grass underfoot. Birds gently tweet from within a wispy fog, dulling the sound and leaving only the outlines of trees on view. Take a deep breath of the fresh air. It is a sublime experience. But the countryside was not always like this. Step back a few years, 4.5 billion to be precise, and the scene is very different. The surface of Earth has condensed from the gas disc from which all the planets formed, and you are standing on one of the first volcanic land-masses. Underfoot, recently molten rock stretches brown and dry to the horizon. Here and there, steam drifts out of small holes, vents in the ground from which volcanic gases emanate. This is a dead Earth, yet to give birth to the first creatures. And here is another major difference: you're looking through the eyepiece of a respirator that covers your whole face. On the other end of the respirator is an oxygen bottle. Don't take your gas mask off, or you will instantly asphyxiate.

As we turned out of the prison gates I began my story. "So, you've got to imagine being on a hot ball of rock when the Earth formed," I said. "The atmosphere had no oxygen for anything to breathe. The planet was enveloped in the earliest gases that came out of the planet or were left over from when it formed.

"In this earliest state as an incandescent ball of rock, Earth probably had an atmosphere of hydrogen and helium, elements so light that they rapidly took flight and dissipated into space, leaving only the gases that bubbled from the interior of the Earth. They were noxious, quickly forming a thick atmosphere. The air was filled with carbon monoxide, sulfur dioxide, hydrogen sulfide, hydrogen, carbon dioxide, and a smorgasbord of others that today still belch from Earth but at far lower concentrations."

"So deadly, basically," my driver interrupted.

"Yes exactly, completely toxic," I said.

Well, toxic for humans, at least, and for much of the life that inhabits Earth today. But the early planet was not entirely unlivable. Once life did get going in the form of microbes, many used the gases as food. By taking up these gases from the atmosphere, the microbes could get the energy and nutrients they needed in order to grow and divide. They munched hydrogen and carbon dioxide, and emitted methane as waste—the compound we now associate more with farting cows than with primordial Earth. Other microbes ate sulfate minerals and produced hydrogen sulfide gas, which in turn was used by other microbes in their own metabolic processes. In this way, the great cycles of the elements—carbon, sulfur, nitrogen, and others—began to turn, feeding the biosphere. This process, of course, is still with us, carried out by the same sorts of microbes and by their barely modified descendants.

This oxygen-free existence was a hard one. Gases were plentiful, but they didn't yield much energy: these early microbes would have gotten at most a tenth of the energy from their primitive foodstuffs as you and I get from the sandwiches that we metabolize in oxygen. Some forms of food, variations of the element iron, would have yielded a hundredth of the energy. Nevertheless, there was just enough energy to support life in the primordial stages of evolution.

"And then what, did it just stay like that with the same mixture of gases?" my driver asked.

"For a long while," I replied. "Life went on like this for a good billion years, maybe longer, without changing much. A microbial world covered in slime." These oxygen-defying lifeforms, known as anaerobic microbes, lived in the oceans, carved out a living on land, and chomped away at the rocks deep underground. But "then something really remarkable happened," I explained. "A microbe discovered something fantastic. Life discovered how to use water to help power its energy needs."

You see, hidden in water are electrons that life could in theory use to collect energy. But to use them requires some chemical wizardry. First, you need to break up the water molecules to get at those electrons, and that isn't easy. Doing so requires a special catalyst. Then the electrons themselves have to be powered up using sunlight. It took some genetic mixing and matching to put together the necessary chemical pathways to make use of electrons that would be quite feeble in the absence of solar power. That could explain why life didn't advance much on this problem for a billion years—it just took a long time for chance to do its thing and find the right biochemical magic to spring forth the water-using sunlight-powered bacteria. The new creatures were the first photosynthesizers that produced oxygen. They were able to use sunlight and water to power their growth and multiplication across the planet.

My driver listened attentively. "And why the water?" he asked. "I suppose it's everywhere, right, it's very easy to get hold of it?"

"And the sunlight as well," I continued. As long as you remain on the surface of the planet, sunlight is everywhere. As for water, about three-quarters of the planet is covered in the stuff; all those oceans, lakes, rivers, and ponds ensure that water is much more prolific than the bubbles of hydrogen sulfide or hydrogen to which life was previously limited. These ancient sources of energy weren't exactly rare, but to get them you needed to be near a volcanic pool or in a deep vent where the gases seeped out. Otherwise, someone might eat it before you. But water was everywhere.

Immediately after the first photosynthesizer discovered its special trick, there was no competition. The gas-eating microbes persisted but could hardly outdo an organism for which the whole world was a free lunch. Very quickly these novel microbes, the cyanobacteria, would spread into every body of water, setting them on a path to evolutionary greatness. Cyanobacteria went on to form an alliance with other cells, which engulfed them and formed algae. In time algae evolved into plants, eventually turning into roses and raspberries and conquering the landmasses. Every green, sunlight-powered creature you see on land and in the sea owes its existence to the discovery billions of years ago that water could be a source of electrons. However, let's back up to the early days, when cyanobacteria were just another of the single-celled creatures doing their thing on primitive Earth—multiplying, replicating, metabolizing.

"Here's the kicker," I told my driver. "These newfangled creatures weren't just another microbe. In breaking up water and gathering sunlight to feed their needs, they produced, like all of us, waste. And that waste was oxygen gas. In little pockets of lakes and at the surface of the oceans, the gas began to accumulate, the oxygen we know today."

This process of accumulation took a long time, in part because oxygen gas had a habit of vanishing. In the early atmosphere, still full of reactive volcanic compounds, oxygen didn't last long: it would react with other gases like methane and hydrogen, and so be removed from the atmosphere. Even the iron in the oceans had a habit of scavenging oxygen gas. The oxygen-emitting microbes went about their business with little influence on the world around them.

My taxi journey this day was itself like a tour back in time to these early events. Every mile on the way home was about a billion years of Earth history: as we turned out of the prison gates, Earth had just formed; on Gorgie Road, the microbes had figured out how to split water; when we reached Haymarket, the great rise in oxygen was occurring, our planet was transforming into one fit for animals.

But are these fairy tales? How do I know that the stories I told my driver are true? The answer is time travel—sort of. No, I don't have a real time machine, but geologists can time-travel in a roundabout way. They can dig up rocks and see what sort of minerals existed on the bubbling, cratered early Earth. These minerals contain clues about what gases were in the atmosphere, because minerals behave in different ways depending on the gases to which they're exposed. For instance, when rocks are exposed to oxygen, they tend to form what are called oxides. Basically, the rocks rust. Oxygen has a voracious appetite for rocks, altering them much as it does the metal in your bicycle.

Now the time-travel factor. Over the ages, the oxidized minerals get buried under the shifting sands of Earth's surface. When, billions of years later, a geologist digs them up, the rocks become time capsules. They can be investigated to see what kinds of gases surrounded them long ago, teaching us a great deal about what was going on around our ancient planet. One thing we've learned is that, before about 2.5 billion years ago, these oxidized minerals were nowhere to be found in any great abundance. Instead, common minerals of the time were exactly those you'd expect in an oxygen-poor atmosphere. In other words, in the samples we dig from deep underground—samples known, in this case, as proxies—we find very few oxidized rocks greater than 2.5 billion years old, but many more after that cutoff. From this we know that early Earth was essentially devoid of the oxygen that we now take for granted.

That oxygen appeared later, thanks to another major development. Recall that, as the cyanobacteria produced oxygen, the atmosphere and the ocean's iron mopped up the gas. Eventually, however, the reactive, oxygen-binding chemicals got used up; they could no longer grab all of the oxygen for themselves. So the leftovers started to accumulate in the air. The oblivious cyanobacteria kept churning out the gas, no matter where it wound up. Soon enough, large quantities of oxygen built up in the atmosphere.

Now, I don't want to leave the impression that the story is quite so simple. If it were, then our geological evidence would look rather different than it actually does. Were there no more to this history than what I've outlined above, then we would see a very gradual oxygenation of the atmosphere as oxygen-fixing reactions slowed down and microbes steadily fed more of the gas into the air. But this is not what the proxies indicate. Instead the rise in oxygen was apparently rapid, at least in the geological scheme of things. Quickly, Earth transformed from a planet with scant oxygen to one with maybe a few tenths of a percent of present-day levels. That is a big shift all at once; something radical must have happened, tipping the balance.

Exactly what this accelerant was is a matter of debate. Nevertheless, some switch was indeed flipped, and about 2.5 billion years ago our planet made a definitive transition toward oxygen-richness. This new condition prevailed for 1.8 billion years or so. And then, another great oxygen infusion occurred. About 700 million years ago, oxygen levels rose relatively suddenly to levels near those we experience today. Again, the reasons for this abrupt shift are not fully understood. But modern Earth had arrived.

"I see," said my driver, "so that's where the oxygen gas came from. And then I suppose the animals, you and me, could use it like we do today. I get it now."

"What's interesting," I explained, "is that oxygen isn't just any old gas. The reason why we can use it to power all the things we do is because it's a very powerful oxidant. You know this every time you light a bonfire or fire up a barbecue. Those old newspapers and coals set on fire and release energy using oxygen as the oxidant. Life does this same reaction."

When I say that life does the same reaction as a barbecue, I mean this literally. Your body carries out exactly the same chemical transformation as a campfire does, burning organic material—the food you eat, whether pastrami or pickles—in oxygen. The big difference between your body and a wood-fire is that, within your cells, the reaction occurs under controlled conditions. If the reaction were not under lock and key, you would spontaneously combust.

"And that gave us lots of energy," my driver concluded with the punch line.

"That's exactly right," I said. "As your bonfire will prove, burning things in oxygen releases impressive amounts of energy. When life figured out how to do this, it had at its disposal a reaction that would produce prodigious amounts, more than those feeble gases and rocks of early Earth. The cyanobacteria, by spewing out oxygen as a waste product of their amazing discovery of water splitting, had spawned an energy revolution."

The consequences were stupendous, because the new oxidant allowed for a huge expansion of life. What's probably most important is that the new energy source allowed things to get bigger. Cells could start to cooperate and form larger structures. The rise of animal life, and the multicellular creatures that eventually produced you and me, begins to become obvious in fossils about 550 million years old. A lot of people think that this efflorescence of creatures, eventually including animals with skeletons, is linked to the rise of oxygen that occurred shortly before this (again, shortly in the geologic sense).

The relationship between size and evolution is very tight. Greater size means novelty—new capacities, new opportunities for interaction with the surrounding ecology, including other animals within it. Importantly, as animals grew in size, they could eat other animals. The animals at the receiving end of this rather unpleasant behavior expanded in stature because doing so helped to ensure they wouldn't become someone else's lunch. Instead, bigger animals could go on to reproduce and propagate their genes. Oxygen kicked off a size-and-complexity arms race.

The result was the so-called Cambrian explosion, which refers to a period in the geological record when suddenly we find many fossils of complex animals. The Cambrian is often confused with the beginning of animal life, but strange pancake- and frond-like creatures appear in the fossil record of the preceding Ediacaran period. Still, if the Cambrian was not the animal's debut, it was a time of significant evolutionary developments—among them, increasing size and animals with skeletons. It happens that skeletons preserve well in rocks, hence the apparent "explosion" in the number of animals.

Not only did the animals get bigger during the Cambrian explosion, but the more they accumulated energy, the longer their food chains became. One animal would eat another and itself be consumed, its own predator becoming prey to another. These food chains became more complicated, more energy intensive, and more widespread. Indeed, a chain has never been precisely the right metaphor, because the relationships of dependency among living creatures are not, and never have been, totally hierarchical, with stronger and more complex animals simply preying on those a rung below. Rather, what emerged in the Cambrian were many intersecting webs of life. Thus in a brief moment of transformation, a biosphere that for billions of years contained only microbial slime gave way to a profusion of living things whose descendants decorate the world we know today. Oxygen enabled dogs and dragonflies, anteaters and aardvarks.

And so, as we turned into Bruntsfield Place, about twenty minutes after leaving Edinburgh prison, the second rise of the gas had occurred and animals took over the Earth, shimmering in the waters and crawling indefatigably onto land. "I'm assuming that this was necessary for us as well," my driver suggested, "we need lots of energy, so the rise of oxygen allowed for us."

"Our brains are wrapped up in all this," I affirmed. "Your brain needs about twenty-five watts to run, less than a conventional light bulb. I like to remind my students that, in the end, we are all quite dim. Anyway, your body needs about seventy-five watts to run, jump, and skip. So, to be an intelligence capable of building spaceships and watching internet cat videos takes about a hundred watts of power. This may not seem like much compared to the needs of a modern home, but for living things, that's not a small amount. Oxygen could allow for that."

So oxygen allows us to be energy-intensive creatures, but is it necessary? That is the big question: Might the profusion of life, and the emergence of intelligence, have been possible without oxygen? Maybe the arrival of animals in the aftermath of the second oxygen boom was a coincidence: it is not out of the question that a complicated biosphere could emerge without oxygen. It is rather difficult to imagine, though. For a start, if you snack on rocks, you have to go finding them all the time, which is very inconvenient and restricts the range of places and habitats you can live in. Oxygen, by contrast, is almost everywhere in the air. You can just breathe it in wherever you happen to be. There are other gases that might do, like hydrogen sulfide; but they offer far less energy than does oxygen. That means either that you can't do as much, or, if you insist on running a 25-watt brain, you'll have to invest a lot of time in eating. Spending all day grazing is not terribly efficient, though. Hunting, cultivating plants, and feeding takes enough time as it is.

Although we cannot say this with total certainty, it does seem that the rise of animals and ultimately intelligence depended on the rise of oxygen. That is the soundest hypothesis we have. If correct, it might also explain why the world remained locked into a microbial way of life for so long. After all, if animals were just the inevitable follow-on from microbes, and lack of oxygen was no constraint, why didn't animals appear billions of years earlier than they actually did? The eons of nothing but microbes suggest that something was holding life back. The rise of oxygen as the trigger for life's complexity revolution makes energetic sense.

Interestingly, the second oxygen boom, 700 million years ago and preceding the emergence of larger and more capable animals, was not the last. It appears that, about 350 million years ago, oxygen's slice of the atmospheric pie rose to about 35 percent, before dropping back to near present-day levels about 100 million years later.

At this point you are probably thinking that an era of even greater oxygenation would produce even bigger animals: with more oxygen in the air, there was even more energy to be had. This could be true of animals, such as most insects, that depend on diffusion to obtain oxygen. Because most insects can't actively pump oxygen deep into their bodies—there are some insects, like cockroaches, that articulate their abdomens in a pumping action—they depend on the gas seeping through tiny channels and into the deepest recesses of their anatomy. Increase the amount of oxygen in the atmosphere, and maybe that oxygen could diffuse farther into their bodies before it was used up, and so insect bodies could grow larger.

Indeed, there is evidence that supports this case. The fossil record features bloated 300-million-year-old insects, among them enormous dragonflies. The extinct *Meganeura* had a wingspan well over a meter and would have been a formidable predator, swooping into the undergrowth of the great Carboniferous forests and snatching other insects, maybe even the earliest four-legged reptiles. Outlandish millipedes and centipedes more than a meter long also roamed Earth at this time, their gargantuan trains of legs rustling across the ancient forest floor in search of prey.

Were these mighty creatures the children of an oxygen binge, a whiff of gas that allowed them to grow to Godzilla-like proportions? Intuitively, this seems likely, although some scientists have their doubts. More oxygen would have supplied more energy, but it also would have created more damaging free radicals—those reactive forms of oxygen atoms and molecules that can tear apart the essential molecules of life. One could argue that adding lots of oxygen to the atmosphere would have encouraged passive users like insects to become smaller, not larger.

Sometimes it is hard to let go of a good story, and no doubt the oversized oxygen-powered dragonflies have a certain appeal. Whatever the truth, scientists do on the whole believe that oxygen plays a central role in explaining how life on Earth evolved. Oxygen keeps appearing in the lineup of culprits; it is always there at the right time. In all these tales of life, who was seen on the night of the crime? Oxygen!

There's an alien story in all this. This obsession with oxygen explains why astronomers have a particular interest in looking for the gas on other planets. If we could find oxygen in the atmospheres of exoplanets, and this oxygen didn't arise from geological processes, we would have a smoking gun for evolving life. The presence of oxygen wouldn't prove the existence of animal life or even intelligence, because an oxygen-rich planet could still be stuck where our own planet was before the profusion of complex life. However, exoplanet oxygen indicates a world with the potential for animals and brains. If we found many planets with large quantities of oxygen, that might indicate a high chance of finding one or two that sport an earthlike biosphere, home perhaps to intelligence. If we find very few exoplanets with large quantities of oxygen, then we will have reason to suspect that oxygen-using intelligence is rare.

By the time we reached my house and I had paid the fare, our march across time was over. Earth had its modern atmosphere, the one we rely on. I stepped out of the cab, thanked my driver for his company, took a delicious breath of the cold Cenozoic air and went on my way.

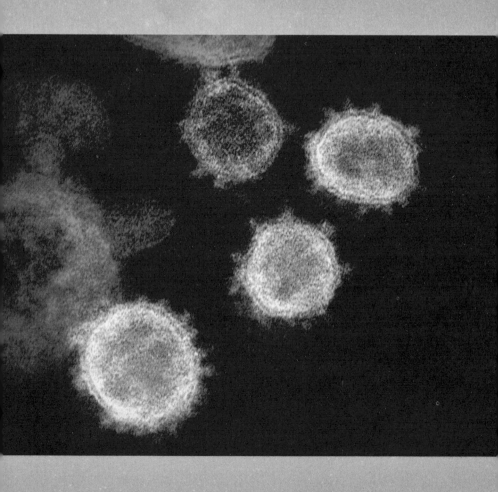

On its own, the SARS-CoV2 virus is a harmless, inert clump of molecules about 100 nanometers across. Place it inside a cell and it can replicate, causing a global pandemic. Is this virus an example of life, or is it something else?

17

What Is the Meaning of Life?

A taxi ride to Haymarket, to catch a train to Glasgow for a talk about prison education.

✳

Few things stir wonder in us quite like space exploration. From Neil Armstrong's moonwalk, which must seem like ancient history to children today, to the adventures of Martian rovers, there is plenty to inspire even those who don't know much about the scientific purposes of space research. The immensity of the universe, the possibility that there are life-forms out there, and the excitement of a human future beyond Earth grip the minds of people who may otherwise share little in common. There really is something for everyone in space, even if that is just entertainment.

It was with this in mind that, in 2016, I started Life Beyond, the prisoner-education project briefly mentioned in the last chapter. Working with the Scottish Prison Service and Fife College, Life Beyond puts prisoners in the shoes of tomorrow's space settlers. The participants design stations for the Moon and Mars, an initiative that engages them in science, art, and many other interests and vocations. The program members have painted station models, written imaginary emails from Mars, and composed lunar blues music. In the process, the prison population are contributing to our ambitious plans to establish ourselves on other planets. Two books have been published with their designs, and their work has won national awards and the praise of astronauts. For me, working with the prisoners has been immensely fulfilling, an opportunity to leave behind the expectations of academia and work on behalf of people who would otherwise have no part in my professional life.

I give lots of talks about science, but on this day, I would be covering the prison work. I was headed to Glasgow to discuss Life Beyond with colleagues who were interested in supporting the project. I got into the back of a black cab and immediately encountered a loquacious character. That comes with the territory: some drivers are stone silent, but many love to chat. This one, an enthusiastic man probably in his forties, started in right away. "It's a funny old world," he said. "I drove this woman this morning who was attending Buddhist classes. She was telling me about how animals have spirits; we're all reincarnated and so there's no point in it all, apart from waiting for the next life of course."

It wasn't an irrelevant thought for my day. What is the point? Why was I designing Mars stations in prisons? Why did I think anyone should be doing this? Why did I think that going to Mars was a worthwhile objective for our civilization? I didn't have any special purpose in becoming so interested in space exploration in prisons; for some reason, I felt it was a worthwhile use of my time. It is my long-held belief that life itself cannot be said to have purpose. The cycle of reproduction and variation, the journey of evolution—it just happens and here we are, on this roller-coaster of mutation. It just is. I wondered what my taxi driver thought.

"So what do you think?" I asked, "what do you think this life is all about?"

"It depends what you mean by life, mate. I mean what actually are we?" he replied. His answer had a straightforward profundity. What do we actually *mean* by the word "life"? This question opens up a fascinating subject.

"Life" is one of those words with a thousand meanings. It has vexed the human mind for millennia. All of us wrestle with the question of what our lives are for. In a quotidian sense, we have to answer this question if we wish to earn a living. Large questions of purpose are reduced to something more concrete. What job should I do? Where should I live? These are the manifestations of the word "life" brought into view by our everyday experiences of it. They are the prosaic embodiment of the urge and need to establish the meaning of existence.

Tugging relentlessly like an ocean current beneath these everyday concerns is some deeper meaning of the word. What is the purpose of life in general? Are we passengers on a tide of fate decided by the cold, uncalculating universe or an omniscient God? Or is there nothing but the blind determinism of evolution operating day in and day out without intention or direction? If we accept a purely deterministic undercurrent to our existence, can we bring meaning into our lives as individuals and as a civilization by creating and believing in our own purposes?

But I don't think it was any of these versions of life that my taxi driver had in mind when he asked me what it was.

What he was thinking about was another interesting question about worms, snails, leopards, and people. He was thinking about the physical matter we call life. What is all this stuff? What is the difference between a creature and a mere object, such that one is living and the other is not?

Like a submerged reef, this question, too, has shipwrecked the minds of thinkers since time immemorial. No matter how completely one tries to focus on physical reality, no matter how objective and reductionist one hopes to be, few human beings can escape the sense that life is imbued with something that differentiates it from a table or chair. What is it, this essence that actualizes and energizes the objects we like to think are living?

Before there was any knowledge of atoms and elements as we think of them today, the ancient Greeks felt sure that life had a special ingredient. This was easy to explain by invoking theories on how the universe was constructed. Attempting to describe why different objects were made of different materials, the fifth century BCE philosopher Empedocles contrived the ingenious idea that everything is made of four substances: air, water, earth, and crucially fire. By mixing these four entities together, different types of material, from oceans and land to tables and carts, could be made. Life was no mystery: an extra dollop of fire caused its energetic and unpredictable temperament.

Aristotle had a similar idea. He thought that everything in the universe contained a substance called matter. He was on the right track. His notion of matter is directly comparable to our own ideas about matter in the modern age. But mixed in with this material was another mysterious substance called form, and this form was made up of soul. The soul was the stuff that made matter think. If you had only a little of it, you would be a plant; a little more and you could be an animal. And the most generous portion was reserved for humans, who had consciousness. Fundamental to both perspectives— those of Aristotle and Empedocles, and of many others besides—is the unshakable conviction that there is some categorical difference between life and nonlife.

In the seventeenth century, however, it became increasingly clear that the differences between life and nonlife, if there were any, were not so fundamental after all. This was the time when chemists began to experiment more carefully and systematically, and through these efforts the properties of the elements came into view. By crushing and squashing, heating and cooling, irradiating and reacting, centuries of effort made it abundantly clear that dogs were made of the same stuff as tables. Carbon, hydrogen, oxygen, and so on. Everything in our world was made of the same atoms and same subatomic particles. No glint of light could be seen between the living and dead, not even a fracture line. No fire atoms or soul appeared in the chemists' test-tubes. This was all rather inconvenient.

To escape the banality of matter, something new was needed. Those who craved an exalted place for life reconfigured Aristotle's soul as an élan vital that would save us and our fellow creatures from the ignominy of being just another agglomeration of the atoms organized in the periodic table. There was no shortage of crackpots and serious scientists willing to proffer theories of what this unusual ingredient was. Perhaps a form of electricity sparked life into existence. Wild and wonderful experiments attaching animal organs to newfangled electrical gadgets would allow us to discover the nature of this vital force, or so it was thought. The ocean of conjectures, however, yielded no catch. Not a single experiment found anything in life that was absent from the nonliving.

But the quest to find some categorical division between life and nonlife survived the failure of vitalism, as researchers turned to the behavior of living things: surely here was something that might characterize life as distinct from the nonliving. Take a look at a dog walking along the road and ask yourself: What is it about that creature that I think makes it alive? Forget for a moment any religious proclivities you might have and steer clear of any metaphysical nuance and ask yourself that question in a very practical way. You might reply with a few things. For one thing, you could observe that a dog is complicated and unpredictable in its behavior. That does seem different from a table, which sits silent and motionless, only changing through the ravages of use and decay. For another, you might say that the dog can reproduce. Baby tables sprung from the union of two tables would be the stuff of a Halloween horror movie. And the puppy that emerges from a canine union grows from tottering furry ball to confident adult dog. Adding a leaf can expand your table, but on its own, it doesn't grow.

But once you interrogate your list of qualities that separate dogs from tables, you will find that many of these supposed distinctions do not map easily to all comparisons between life and nonliving things. A tornado, hurtling and twisting through a neighborhood, snakelike in a whirl of rooftops and detritus, is no more predictable than a dog: complex behavior is not the sole domain of life. And if tables don't reproduce, some inanimate objects do multiply. Crystals growing in a bath of chemical feedstock might break apart and continue to exist separately, a primitive form of behavior that looks a little like reproduction. And if you leave them in their liquid bath, they too can grow, expanding from tiny nuclei to fist-sized agglomerations.

We can carry on like this, listing every feature of life in an attempt to identify that special essence of the living form, but there is not one that isn't subject to exceptions. Even metabolism, squarely in the territory of biochemistry, is not limited to life. As we saw, there is little difference between "burning" a sandwich for energy and burning down a forest. Forest fires burn organic material, such as trees, in oxygen and emit carbon dioxide and water as waste products. Exactly the same chemical reaction goes on in our bodies, albeit within the confines of living cells.

Evolution might seem a last redoubt. This inexorable process of mutation and selection seems unique to life. After all, there is no evolution without a genetic code. But it turns out that evolution is not confined to ecology or even biology. Some researchers have made molecules evolve in the lab. Computer software has been made to evolve in rudimentary ways a little like a genetic code, although some people would say that these programs are the products of a mind that evolved, so they aren't a clear-cut case of evolution in the nonliving world.

The obverse of this coin is also instructive: many things that we consider alive do not exhibit characteristics we expect of living things. Your coffee table cannot reproduce, and that seems a sound basis on which to reject it from the club of living beings. But then, a mule can't reproduce either. If you saw a mule dragging a cart along a dusty road, though, you'd be hard-pressed to defend the position that it isn't alive, simply because it is sterile from birth. And what about you and me and rabbits? We can't reproduce, either. We require mates to make that possible. Is a lone rabbit leaping across a field dead, becoming alive only when it has found its happy union? We have descended into absurdity and in the process lost some more ground in the battle to declare the uniqueness of life.

The aforementioned physicist Erwin Schrödinger, famous for his feline thought experiment, waded into this fray without much success. Schrödinger grappled with the idea that life extracts energy from the environment to make order out of the universe—to assemble itself, to turn atoms into puppies and trees using energy that, on our world, comes mainly from the sun. Yet, while life certainly funnels the universe's energy into the assembly of complex machines, it is not alone in this regard. Complexity emerges in the eddies and swirls in your coffee cup and in the turbulent globules of gas at the surface of the sun. As energy is dissipated in the universe, so temporary complexity emerges in the process—a kitten here, a vast and beautiful solar magnetic distortion there. The universal phenomenon of emergent complexity may be given exquisite form in living creatures, but the same physics is at work in many inanimate things. Schrödinger and his followers who explored emergent complexity could find no throne for life.

Do you see in all this a certain desperation? A forlorn attempt to find something, just anything, that will make the living creature different from the nonliving? If you do, you are not alone. I join you, as do many others.

The problem with defining life in a way that categorically distinguishes it from nonlife is that life isn't just out there, waiting for us to hold it. As far as we can tell, there isn't a physical thing you can point to and say, "There, that's life." Rather, life is a property of objects that we intuit, and different people have claimed to find this property in different objects. Humans invented this term because it is useful, but its precise contours have never been established. This makes life different from certain things that are more easily defined. Take, for instance, gold. You can tell me just what it is. Fine, you might not be able to off the top of your head, but if you search the internet, you can soon furnish me with information that tells me exactly what substance gold is. You could give me its boiling point, its atomic number, its electron structure, and more. Philosophers call stuff like gold a "natural kind," an object or substance whose characteristics are rooted in basic physical properties that we can define and enumerate precisely. Life isn't a natural kind.

At least, it isn't for the moment. Some people think we will be able, eventually, to define life in the way we can define gold. After all, gold, too, has not always been a natural kind. (Well, that depends on which philosopher you ask. But let's move on.) If you had asked Aristotle what gold was, he would have muttered something about matter and form, perhaps thrown in something about the soul and atoms of fire, and left you feeling no less bewildered than you were before you sat down in the Athenian sunshine to ask his advice. But times changed; eventually chemists prodded and poked enough gold with enough care to know how this material was constructed. Now, with this hard-won knowledge, we can state the exact nature of gold. Could the same be true of life? Perhaps with enough progress in biology and physics, you will one day be able to tell me exactly what life is, providing a definition that admits no exceptions and submits to no counterargument.

However, there is another possibility. We might never be able to define life. What would happen if instead of asking you to define gold, I asked you instead to provide me with a definition of a table. You might respond with something like, "Well, it's a bit of furniture I put things on." Then I would show you a picture of a stool, which looks remarkably like a small table, and give you a quizzical sideways look. "Mmmm," you might reply. Yes indeed, "Mmmm." And so we could go on for hours arguing about what a table is, probably sidewinding into arguments about what a stool and a chair are and whether sitting on a coffee table turns it into a chair. Why would our conversation be such a dead end? The reason is simple. The word "table" does not refer to some type of element or atomic structure like gold; rather, it's just a word that humans dreamed up to refer to objects having certain qualities in common. Most of these objects, these tables, quite clearly are things we all recognize, even though there may be a great deal of difference between your kitchen table and the queen's state banquet table. But rummage around at the edges and we can find plenty of objects that show, with alarming simplicity, that the word "table" is a linguistic convenience encompassing much mystery and contradiction. Try to define a table in physical terms, separately from the intentions you have for its use, and you cannot exclude a stool.

Perhaps, like a table, life is a nonnatural kind. It exists in our ever-inadequate definitions, but, from the perspective of the physical world, there isn't really a division between life

and nonlife. Instead it can be more useful to think of a grada-
tion between a simple molecule and a human. As you increase the complexity of matter, so certain features of life
begin to kick in. But it is not necessary that something possess all these features at the same time or with the same intensity in order to be alive. Hence our mule that hasn't got
reproduction but has some other features of the living. In this
view of chemical complexity, where you draw a line between
life and nonlife is arbitrary and depends on the list of things
you've decided you want to be part of your definition of life
in the first place. Our definitions are not therefore useless,
just partial. Life encompasses some matter that does some
interesting things such as growing, reproducing, evolving,
metabolizing, and otherwise exhibiting complexity. This material is of special interest to us, not least because we belong
to it, and so we like to cordon it off from other matter and
declare it unique. But look at life's edges and we find it seamlessly grading into other matter, depending on the word
games we play.

The putative division between life and nonlife is hopelessly
encouraged not just by our desire for semantic clarity but also
by religious and moral arguments that presume a special
place for humanity. The acceptance that life is a nonnatural
kind—a mere word that places an artificial, mutable, and very
permeable border around some particularly interesting lumps
of organic material—faces strong resistance from those who
fear that such a posture leads to nihilism. We balk at exploring the borders of life because we don't want to find that
our privilege is undeserved. This alone may motivate the age-
old interest in finding a definition of life, drawing the line

that none can cross so that finally we may enjoy our hallowed sphere, shielded from ambiguity.

But what do we gain from this obsession with articulating boundaries between life and nonlife? Consider a world in which humans are just complex organic chemistry. In this world, "life" is a useful way of roughly hiving off the part of chemistry that defines the work of biology. Would this be so bad? Fears of nihilism notwithstanding, I see no reason why anyone's moral compass would go haywire. Does the definition of a single word erode your capacity for empathy for other lumps of organic matter that share similar features, other lumps of matter that for convenience share the same label? Indeed, one might convincingly argue that embracing life as a nonnatural kind would enlarge our sphere of empathy. Perhaps those organic forms that sit at the boundaries of the word life also merit care. A firm definition of life could leave a mule to be treated like a table. By contrast, accepting that humans are mere chemistry, and life merely an operationally useful word, humbles us. It is potentially a route to circumspection and thoughtfulness, not the wantonness that currently is displayed toward anything not deemed alive (as well as much that is).

If science should, one day, take us to a clearer formulation of the things that constitute life, then so be it. With that shift, the boundary will tighten and sharpen. But why push this endeavor for the sake of fulfilling a pathological demand for specialness? We don't actually need life to be like gold.

My own feeling is that it never will be—life will never join gold in the pantheon of natural kinds. Life will forever remain a useful term to talk with. The reason for my conviction lies simply in the reality that life is not, like gold, a single set of atoms arranged in an ordered form that admits of exact definition. Life's capacity for chaos, for emergent properties, and the immense complexity in the way in which atoms can be arranged to generate a vast diversity of matter that does multitudinous things, leads to a feeling that not only will life obviate any possibility of precise definition but also that it might be better that we stop trying to pretend otherwise.

A loose definition of life allows us the possibility of including new forms of matter as our knowledge of the universe develops. Maybe in the distant future, on another planet, human explorers will stumble across some material that exhibits complicated interactions with its environment, maybe even shows what we consider awareness of its surroundings. These qualities would bring it within the purview of material that on Earth we would consider living. At the same time, this thing's composition and complexity may be such that we cannot easily specify whether it is an instance of life in the way we think of life on Earth. Drawing on a hard definition of life, we would exclude it from the purview of living things, enabling a certain recklessness toward it. Perhaps we could even destroy it, just to be safe. A soft definition of life might lead us to other outcomes. We would instead be encouraged to debate and modify our ideas about the living form.

Rejecting the millennia-old pursuit of a definition of life may open our minds. To the scientific thinker who seeks taxonomic clarity, accepting a fuzzy definition of life seems untidy. But maybe it is the more honest approach, and that too should appeal to a scientific thinker, and to everyone. We should not assume that nature made this thing we call life fundamentally distinct from another thing we call nonlife. If we avoid that mistake, we will find ourselves better equipped to learn from all that we find in the universe, those material substances that are part of ourselves and everything around us.

From the standpoint of chemistry and physics, there is nothing exceptional about life on Earth. But how common are living worlds in the galaxy, or planets where creatures build radio telescopes, like the Atacama Large Millimeter/submillimeter Array?

18

Are We Exceptional?

A taxi ride from Mountain View to Sunnyvale, California.

❖

Sometimes profound questions come from trivial beginnings. This was such a time. I was taking a taxi from a motel in Mountain View to a hardware store twenty minutes away in Sunnyvale to pick up a cooler. I needed to preserve samples that we were planning to collect from a space experiment. The samples would arrive at the Port of Los Angeles in just a few days, after returning to Earth from the International Space Station on one of SpaceX's Dragon capsules. Actually, I was getting a little desperate to find the cooler after at least three shops had come up empty. So the last thing on my mind was the meaning of life.

My driver had asked me my profession, and I gave her a potted summary. She was a sincere and quite intense person. When I mentioned my work on the search for alien life, her own curiosity was piqued.

"I'd be interested to know," she said, looking at me from behind a pair of green, round-rimmed glasses. "I'd really like to know. Is there anything else out there or is it just us? I don't think much about this, but every now and then I do think about it. When you see TV programs on planets, you think, is there anything out there?"

"Does it matter to you whether we are alone or not?" I asked.

"We just want to know. It's not that it affects our lives, but what if we are alone? We could be the only things out there," she said.

There is an inescapable urge, buried deep in the human psyche, to be exceptional. I think this word—exceptional— must be one of the most confused in the English vocabulary, yet we seek to know whether it applies to us.

"So would that make us more special, if we are alone?" I asked.

She thought a bit and then continued. "It doesn't change whether I'm special to people," she said. "But it's a big question."

I sat quietly, looking out of the window. Whether human beings are exceptional in the universe is a question that drives at the heart of our hopes and anxieties. For many people, to be unremarkable would be to deny any purpose in human life. A universe in which we are not special is one in which we are demoted to the status of mere animals, as some people see it. So it is perhaps not surprising that when you are sitting in a taxi talking about alien life, you start to think about what all this means for us. Are we anything worthy in this grand drama, and does the answer to that question depend on whether we are the only intelligent creatures in the cosmos? Needless to say, there is no simple answer. This is one of those questions that includes many more within. What exactly does it mean to be exceptional? What is it that's exceptional—

individual humans, the human species, planet Earth, or something else entirely?

As a scientist, I'm going to pick up on my taxi driver's question from a purely scientific angle. What I mean by this is that I am not going to entertain a discussion of whether you as an individual are exceptional. From a purely factual standpoint, the answer is obvious and therefore the question uninteresting: no individual is identical to any other, so from that elementary perspective, you are exceptional. If instead to be exceptional is to be admirable, I will leave that for others to judge.

Throughout this book, I have taken up a different question of exceptionalism, one amenable to scientific investigation: Is the very existence of life on Earth exceptional? We don't know. We know that the molecules from which we are constructed are made up of simple bits and pieces that probably rained down on the primitive Earth and also were made on Earth itself. Yet we don't know if this is all it takes to make life—whether the assembly of these bits into replicating cells was inevitable. Our search for life beyond Earth might bring some resolution on whether the emergence of life on a planet like Earth is exceptional or common. That search can also help us figure out whether intelligence is likely to emerge once there are cells, granted that the gap between these is enormous and may persist for billions of years. Intelligence may be common or rare, or it may be the singular capacity of humans.

We have firm answers to one version of the exceptionalism question, namely whether our planet is unique. Indeed, our answers are so firm that few think to ask. But it was not always this way. The ancient Greeks were divided: some believed that Earth was not exceptional, that there might be other heavenly bodies similar to it across the heavens. But it was Aristotle whose view prevailed into the Middle Ages. Aristotle asserted that Earth was the center of the universe, and the sun revolved around it. This position was attractive to later monotheistic religions, which placed our planet, and humanity in particular, at the heart of God's heavenly design. For more than a thousand years, our exceptional place in the universe was unquestioned. It took the heresy of Nicolaus Copernicus and his 1543 book *De revolutionibus orbium coelestium* (*On the Revolutions of the Heavenly Spheres*) to demystify Earth.

With every generation since then, our Earth has lost more of its exceptional character. After Copernicus, Earth may have been the slave of the sun, but the solar system itself could nonetheless be the work of a creator who provided us life-giving warmth. Yet as we peered farther out into space, it became painfully apparent that many of those tiny white dots in the night sky were themselves suns. We didn't know much about them, but it was impossible to ignore the realistic possibility that orbiting them might be other worlds like ours. With further improvements in observation, we could see that these suns themselves were orbiting something else. Great clusters of stars moved in regular fashion around the indefinable centers of what became known as galaxies. Soon enough we understood that galaxies contained vast quantities of stars and that the universe itself was full of these galaxies. Billions of suns in billions of galaxies. This was, seemingly, the completion of the Copernican revolution: no one could believe that, among the trillions upon trillions of planets in the universe, Earth is special. Statistically, earth-like planets may be unusual, but a small percentage of a very big number is still a big number.

Yet now in the twenty-first century, something extraordinary, almost perplexing, is underway. In the last few decades, humans have been learning that, while there are many stars like ours out there, planetary systems vary greatly. The exoplanet search has not found replicas of our solar system, facsimiles of the same process of planet formation. So far, every system we have scrutinized is unique, not merely in the spacing and architecture of the systems themselves, but also in the worlds they harbor. Puffy planets, super Neptunes, hot Jupiters, ocean worlds, rocky planets made of carbides—no end of oddball descriptions litter the scientific literature.

The rocky planets, those most like Earth, are subject to a remarkable range of permutations. Some are tidally locked to small red dwarf suns, with one side of the planet always pointing toward the star, much as our own Moon always shows us the same face. What are the implications for life on a planet like this, with one side permanently lit and the other in eternal darkness? We don't know. Some rocky exoplanets have highly elliptical orbits, so that they fly in close to their star before spending long periods in the freezing depths of space, creating climates that swing wildly from high heat to deathly frigidity. Other exoplanets are blasted with radiation, and still others orbit stars whose lifetimes are probably too short to provide a nursery for intelligence.

If we were to find a piece of rock with water and appropriate radiation and temperature levels, we may still not be looking at an environment sufficiently earthlike to support life.

Essential to life on Earth is our system of crustal plates that, through their incessant diving and melting into the depths of the planet, cycle the vital elements for life, energizing and fueling the biosphere. It may be that a similar plate system is required everywhere, or at least some of the time. The wrong size of planet or the wrong amount of water, and the plates may seize up, turning the planetary surface into a great slab of unmoving rock, like Mars, or forever submerging the crust under a deep ocean. Life would be relegated to an ocean existence, if there is any at all.

And what of the atmosphere? A planet that is in many ways earthlike could still have too little atmosphere, or too much. The particular concentrations of gases could leave the atmosphere and the surface below too hot or too cold. Even if the star is much like our sun and the orbital distance much like Earth's, the qualities of the atmosphere may be such that the planet receives too much radiation or not enough sunlight, preventing the emergence of life or its subsequent evolution.

What this all adds up to is a new lease on terrestrial exceptionalism. Having found ourselves in a universe of similar suns, it may turn out that the conditions enabling life on Earth still cannot be found anywhere else—at least, anywhere else that we can detect. What an irony it would be if, in challenging Aristotle's claim for exceptionalism, we were to discover that Earth is exceptional after all—a singular concatenation of physical conditions that produced the only way to be alive, whereas there are so many ways for a planet to be dead.

This, basically, is the question we are trying to answer: How many paths are there to life, and from there to intelligence?

How much variation is there in living worlds? Do life and evolution demand a range of planetary conditions so narrow that the natural vicissitudes of planetary formation will nearly always stand in the way, and any successes will look just like Earth's? Or are the tolerances wide enough that many flavors of worlds could host many flavors of biosphere? So far, we are quite understandably looking for planets similar to our own. But maybe, if and when we do find extraterrestrial life, it will be on a world vastly different from Earth. In other words, we are presuming that life is picky. Again, this is understandable; seeking out earthlike worlds makes the search manageable. But not necessarily successful.

Of course, none of this brings us closer to the kinds of answers that religions have provided. If Earth turns out to be unique, or if we find that life exists only on worlds just like Earth, neither discovery will prove the existence of a creator. But perhaps astronomy and creed will converge on the view that Earth really does have a special place in the universe—the only place, or one of the few places, where life can exist and evolve. In this sense, we are learning from exoplanets that the Copernican revolution is far from complete. Five hundred years on, we remain unsure as to whether our world is something unusual and even unique. The difference is that, with modern telescopes, we could actually find out what the truth is. We need not rely on faith to decide whether Earth is exceptional; we could someday have evidence.

There is one aspect of our existence that we can say with certainty is unexceptional: life, where it does exist, joins all other matter in following the laws of physics. At a glance, this point might seem trivial. By definition, physics describes how the matter and energy contained in the universe work. If we find some material or behavior that lies outside our current understanding of physics, that does not place it "beyond" physics; rather, it just means physics must be revised to account for this new finding. What is nontrivial in this observation—that life is bounded by physics—is that the structure and behavior of life are not extraordinary. The emergence of life may be incredibly rare, even unique to Earth, but the way it works is not so remarkable as to elicit great surprise.

Consider the tremendous variety of flying creatures that evolution has produced. *Mellisuga helenae,* a hummingbird that lives only in Cuba, is between 5 and 6 centimeters long and can weigh less than 2 grams, making it the smallest bird on the planet today. Compare that with the extinct reptilian *Quetzalcoatlus,* a leviathan with an 11-meter wingspan, the same size as a Cessna light aircraft. Yet, as different as *Mellisuga helenae* is from an eagle or albatross, to say nothing of bygone predators, these animals all stay airborne the same way. Their bodies obey the laws of aerodynamics, which say that the area of a wing and the speed at which it travels determine how much lift it will generate. A flying animal has no choice but to follow these rules, or else it isn't a flying animal. The shapes of flying creatures are similar because aerodynamics is the same everywhere, not a matter of whim or contingency.

Next time you see a fish dashing through some rocks in the sediment of a stream or river, look at its shape. If it is a fast-moving fish, maybe the sort of animal that has some predators it needs to avoid, its body will have a streamlined, fusiform shape—that is, a structure tapered at both ends. That is the best way to move fast through water. Dolphins have the same design. They may not need fusiform bodies in order to escape predators, but they might find such sleekness useful when catching other fast-moving fish. In some ways, the crudely similar shape of dolphins and fish should surprise you because dolphins are mammals and fish are, well, fish. Why do two very different creatures end up looking the same? What if I told you that the extinct reptilian ichthyosaurs that plied the Mesozoic seas over a hundred million years ago also had streamlined bodies, a little like modern fish? Now we have a third type of creature with the same basic body plan.

I am sure you have already alighted on the reason. This is physics at play. If you want to move fast through a liquid, like an ocean, then a streamlined body is better than a cuboid, flat one. As evolutionary biologists have observed before, if we do eventually find alien fish swimming fast through a distant ocean, they too will be streamlined. The same laws of physics operate across the universe. Physics governs every aspect of life, from the atomic structure of the molecules in living cells to the behaviors of whole families of creatures.

This sort of thing used to be a mystery, creating an opening for some superior intelligence, God or otherwise, whose hand must have been at play in directing the workings of the animals. As long as life's guiding principles were unfathomable, it made sense that there would be a puppeteer pulling the strings. But we now see much more clearly that the forms taken by life, and the activities undertaken by living things, are not so difficult to explain. For instance, we can use physics to describe how a large collection of living things can operate as a unit, without anyone guiding them. The size, shape, and sheer extent of ant nests, which can cover an area the size of a soccer field with intricate tunnels and connecting lanes and walkways, might suggest the work of a hive brain, a queen ant in whose head the design for this entire edifice must surely be lodged, with each and every detail painstakingly forwarded to her workers, each one of which toils on a small part of the empire. But the queen is no architect, poring over plans and overseeing construction. Rather, ants react to each other. The fewer of them are present, the faster they work; then they slow down when there are too many cooks in the kitchen. No one has to tell them what to do: an elementary set of feedback loops, and the exchange of the simplest pieces of information in their chemical pheromones, are all the direction they need to build cities.

These are the laws of physics working through life. From bird flocks to herds of wildebeest, we find the same principles at work, not the will of an awesome power. Nothing lies outside explanation; there is no élan vital. Humans and all other life on Earth and anywhere else in the universe are the organic manifestation of physical equations, mathematics given biological form.

So humans are not exceptional even on Earth, but life on Earth may well be exceptional in the universe. For although the emergence and many paths of life unyieldingly obey the physical laws of the universe, life itself could be unusual. It is matter in the universe, subject to the same boundaries as all other matter in the universe. (At least, all "normal" matter; it could be that dark matter, mentioned in chapter 12, is quite different, though it would hew to other inescapable physical laws.) But life may be a very rare kind of matter. Like a fine cheese made from common ingredients, the finished article itself may be scarce, an unusual twist on that which is ordinary.

In answer to my taxi driver's question, then, I would have to say: it depends. Whether life on Earth is exceptional, and whether humans are the exception within the exceptional, depends on what exactly you are asking about. This isn't merely fence-sitting. I think it is fascinating that aspects of our existence may be quite mundane, the mere products of inevitable physics, yet out of that banality can still emerge uniqueness.

Another response is that it doesn't really matter to us as individuals whether humans are exceptional, or whether life on Earth is. The answer to that question should make no difference in our lives. We have learned that, at the atomic level, nothing whatsoever distinguishes humans from other living beings or, indeed, from rocks hurtling through the cosmos. Yet this incontrovertible fact has had little impact on our values. Maybe it should have greater impact, but, as an empirical matter, it doesn't. Nor do we linger with the reality that our bodies are just like those of so many other creatures in that they are roughly symmetrical with a central axis of rotation and eyes positioned in the direction of motion. Again, this is just physics guiding evolution.

In the everyday here and now, your exceptionalness is determined by how you behave in relation to others and what you give to society. That is in your control. In this endeavor lies the search for individual purpose, and for most of us it will have nothing to do with whether we are alone in the universe. Whether life is unusual is something the scientific method will reveal in due course; whether you as an individual are fulfilled in a way that is felicitous to your fellow human beings is something for you to decide.

As we deepen our quest to discover the nature of life in the universe, we will not only find out much about ourselves, but we also will confront great challenges, from preserving this oasis of living things we call Earth to establishing societies on distant worlds and finding life elsewhere. Yet we should not expect to uncover in these scientific and technical endeavors ultimate purposes for ourselves. The quest to understand life in the universe is itself the purpose. From that purpose will emerge previously unimaginable discoveries that will color and enrich our self-awareness and perception, perhaps altering what life means to us as individuals and changing the trajectory of our civilization in ways we cannot foresee.

further reading

None of the essays in this volume is intended to be an exhaustive discussion of its subject; if they were, the book would be about twenty times longer. Instead, I hope to have introduced readers to important and provocative ideas, about which others have also said much of interest. Those who wish to learn more might turn to the recommendations below, organized thematically by chapter. The works here vary from popular to academic, with a couple of journal articles thrown in. Some of the suggested texts are old because the best writing is not necessarily new—let us recall that civilization did exist before the internet—and because the quest to understand life in the universe has a history worthy of our attention. I have also cited a few of my own items where they have shaped the major points of a chapter.

1. Are There Alien Taxi Drivers?

Simon Conway-Morris, *Life's Solution: Inevitable Humans in a Lonely Universe,* 2003

An exploration of the phenomenon of convergent evolution—the tendency of life-forms to reach common solutions in the face of survival challenges—and its implications for evolutionary outcomes on Earth and possibly elsewhere. A dense, thorough, and important book.

Nick Lane, *Life Ascending: The Ten Great Inventions of Evolution,* 2009

An approachable text, suitable for all readers, about some of the great innovations in the process of evolution.

John Maynard Smith and Eörs Szathmáry, *The Major Transitions in Evolution*, 1995

A rigorous book outlining significant developments in the history of life on Earth, from transformations in the process of genetic transmission to the emergence of language.

2. Would Alien Contact Change Us All?

Michael J. Crowe, *The Extraterrestrial Life Debate 1750–1900: The Idea of a Plurality of Worlds from Kant to Lowell*, 1986

A well-written, scholarly work on the history of our thinking about extraterrestrial life.

Steven J. Dick, *The Biological Universe: The Twentieth-Century Extraterrestrial Life Debate and the Limits of Science*, 1996

A tome detailing long-standing discussions of extraterrestrial life and the worldviews those discussions implicate, with many colorful diversions.

Bernard Le Bovier de Fontenelle, *Conversations on the Plurality of Worlds*, 1686

This old book, discussed in chapter 2, is a delight to read. Modern versions can be found online and in print.

3. Should I Be Worried About a Martian Invasion?

Albert A. Harrison, "Fear, Pandemonium, Equanimity, and Delight: Human Responses to Extra-Terrestrial Life," *Philosophical Transactions of the Royal Society A*, 2011

A scientific paper addressing the various ways in which humans might respond to contact with extraterrestrial intelligence.

Michael Michaud, *Contact with Alien Civilizations: Our Hopes and Fears About Encountering Extraterrestrials*, 2006

A meticulous and thought-provoking book about the potential consequences, good and bad, of alien contact and efforts to achieve it.

4. Should We Solve Problems on Earth Before Exploring Space?

R. Buckminster Fuller, *Operating Manual for Spaceship Earth*, 1969

In his inimitable style, Buckminster Fuller reflects on humanity's developing relationship with Earth's resources and on possibilities for a sustainable future.

Charles S. Cockell, *Space on Earth: Saving Our World by Seeking Others*, 2006

My book for the general reader proposing that environmentalism and space exploration should be seen as pursuits of the same goal: creating sustainable communities in the cosmos.

Douglas Palmer, *The Complete Earth: A Satellite Portrait of the Planet*, 2006

A beautiful collection of images illustrating how satellites can help us appreciate the grandeur of our life-bearing planet.

5. Will I Go on a Trip to Mars?

Rod Pyle, *Space 2.0: How Private Spaceflight, a Resurgent NASA, and International Partners Are Creating a New Space*, 2019

Pyle brings us up to speed on private-sector and government efforts to make space more accessible.

Wendy N. Whitman Cobb, *Privatizing Peace: How Commerce Can Reduce Conflict in Space*, 2020

Another valuable book looking at the changing paradigm of space travel in an era when private exploration is becoming possible.

Robert Zubrin and Richard Wagner, *The Case for Mars: The Plan to Settle the Red Planet and Why We Must*, 1996

A classic general-readership book arguing for the exploration and settlement of Mars.

6. Is There Still Glory in Exploration?

Buzz Aldrin and Ken Abraham, *Magnificent Desolation: The Long Journey Home from the Moon,* 2009

A personal tale that captures the broader appeal of exploration, through moonwalker Buzz Aldrin's own experience.

Charles S. Cockell, "The Unsupported Transpolar Assault on the Martian Geographic North Pole," *Journal of the British Interplanetary Society,* 2005

My paper envisioning a possible overland expedition to the Martian North Pole from the edge of the polar cap, with details of the route explorers might follow, challenges they will face, and means of preparation.

Leonard David, *Mars: Our Future on the Red Planet,* 2016

An accessible discussion of long-term plans for the exploration of Mars.

7. Is Mars Our Planet B?

Mike Berners-Lee, *There Is No Planet B: A Handbook for the Make or Break Years,* 2019

Without repudiating space exploration, Berners-Lee grapples with some of the major environmental challenges we face on Earth, the planet most suitable for humans.

Stephen Petranek, *How We'll Live on Mars,* 2015

A brief, easy read on some of the obstacles to living on Mars.

Christopher Wanjek, *Spacefarers: How Humans Will Settle the Moon, Mars, and Beyond,* 2020

A wonderful book, packed with information about long-term plans for space settlement and how it can be achieved.

8. Do Ghosts Exist?

Jack Challoner, *The Atom: A Visual Tour*, 2018

A beautifully illustrated guide to the structure of the atom and the history of its discovery.

Lisa Randall, *Dark Matter and the Dinosaurs: The Astounding Interconnectedness of the Universe*, 2015

An absorbing popular read about the nature of matter and the universe.

9. Are We Exhibits in an Alien Zoo?

Stephen Webb, *If the Universe Is Teeming with Aliens . . . Where Is Everybody? Seventy-Five Solutions to the Fermi Paradox and the Problem of Extraterrestrial Life*, 2002

Sound responses to the so-called Fermi paradox.

Paul Davies, *The Eerie Silence: Searching for Ourselves in the Universe*, 2010

A popular discussion of the hunt for alien life in the universe and what it means.

10. Will We Understand the Aliens?

Barry Gower, *Scientific Method: A Historical and Philosophical Introduction*, 1996

A fine scholarly account of the history and development of the scientific method.

Thomas S. Kuhn, *The Structure of Scientific Revolutions*, 1962

A classic philosophical case about how scientific change occurs. Kuhn's arguments were transformative and are still debated.

Karl Popper, *Conjectures and Refutations: The Growth of Scientific Knowledge*, 1962

From one of the twentieth century's greatest philosophers of science, a serious treatment of scientific method and scientific knowledge.

11. Might the Universe Be Devoid of Aliens?

Peter D. Ward and Donald Brownlee, *Rare Earth: Why Complex Life Is Uncommon in the Universe*, 1999

A popular account discussing the various features of Earth that might lead us to conclude that complex life, and therefore intelligent life, is rare in the universe.

Duncan Forgan, *Solving Fermi's Paradox*, 2018

More explanations for our failure thus far to observe intelligent aliens.

12. Is Mars an Awful Place to Live?

Charles S. Cockell, "Mars Is an Awful Place to Live," *Interdisciplinary Science Reviews*, 2002

In this paper I argue that Mars will eventually be populated by stations full of scientists, explorers, and others with business on the planet, but not by millions of people enticed by its exotic living conditions.

Robert M. Haberle, et al., *The Climate and Atmosphere of Mars*, 2017

A textbook outlining atmospheric conditions on Mars, valuable for any consideration of settlement challenges.

13. Will Space Be Full of Tyrannies or Free Societies?

Daniel Deudney, *Dark Skies: Space Expansionism, Planetary Geopolitics, and the Ends of Humanity*, 2020

A sobering counterpoint to optimistic views of space exploration and post-Earth enthusiasm.

Everett C. Dolman, *Astropolitik: Classical Geopolitics in the Space Age*, 2001

A theory of cosmic geopolitics, arguing that astrogeography—positions and distances in space—will be essential to the future of security strategy.

14. Do Microbes Deserve Our Protection?

Robin Attfield, *Environmental Ethics: A Very Short Introduction*, 2018

A primer on some of the major concepts in environmental ethics.

Charles S. Cockell, "Environmental Ethics and Size," *Ethics and the Environment*, 2008

In this journal article, I lay out my views on the place of microbes in environmental ethics and consider how the size of creatures influences the protection humans grant them.

Joseph R. DesJardins, *Environmental Ethics: An Introduction to Environmental Philosophy*, 1992 (fifth edition, 2012)

Another useful starting point for those looking to get a handle on this important subject.

15. How Did Life Begin?

David W. Deamer, *Origin of Life: What Everyone Needs to Know*, 2020

As the title suggests, a discussion of the origin of life geared toward general readers.

Robert M. Hazen, *Genesis: The Scientific Quest for Life's Origins*, 2005

Although some areas of research have advanced, Hazen's book remains an accessible discussion of scientific theories of the beginnings of life and of key experiments and observations supporting those theories.

Eric Smith and Harold J. Morowitz, *The Origin and Nature of Life on Earth: The Emergence of the Fourth Geosphere*, 2016

A scholarly text focused on a critical area of research: the coevolution of Earth and life.

16. Why Do We Need Oxygen to Breathe?

Donald E. Canfield, *Oxygen: A Four Billion Year History*, 2013

Canfield investigates the history of oxygen on Earth and discusses decades of scientific findings about the gas's importance with respect to biology.

Nick Lane, *Oxygen: The Molecule that Made the World*, 2002

Another approachable book about the history of oxygen and its relationship with life.

17. What Is the Meaning of Life?

Mark A. Bedau and Carol E. Cleland, *The Nature of Life: Classical and Contemporary Perspectives from Philosophy and Science*, 2010

A textbook collecting scientific and philosophical views, from many eras and disciplines, on what constitutes life.

Paul Nurse, *What Is Life?: Understand Biology in Five Steps*, 2020

A thoroughly readable book from a Nobel Prize winner on the nature of life, its basic machinery, and how we think organisms work at the molecular scale.

Erwin Schrödinger, *What Is Life?*, 1944

Schrödinger's musings on the nature of life, with prescient ideas about the genetic material before the discovery of DNA.

18. Are We Exceptional?

Sean Carroll, *The Big Picture: On the Origins of Life, Meaning, and the Universe Itself*, 2016

A sweeping look at what we know about the universe, from the subatomic scale to the cosmological.

Charles S. Cockell, *The Equations of Life: How Physics Shapes Evolution*, 2018

In this book intended for all audiences, I consider what we know, and what we are learning, about the physical principles that shape life at all levels of its hierarchy, from atoms to collections of organisms.

Viktor E. Frankl, *Man's Search for Meaning*, 1946

Informed by his training as a psychologist, Frankl explores the quest to achieve a purposeful life amid the worst conditions imaginable: a Nazi concentration camp. From a survivor of Auschwitz and Dachau, this remarkable book remains highly influential three-quarters of a century after its initial publication.

Jonathan B. Losos, *Improbable Destinies: Fate, Chance, and the Future of Evolution*, 2017

A brief on behalf of the inevitability of evolution and the possibility that much of biology is unexceptional—that evolutionary outcomes typically are preordained by structural factors.

Aleksandr Solzhenitsyn, *The Gulag Archipelago*, 1973

According to physics, none of us is exceptional. But if we take from this a nihilistic lesson, the result is likely to be disaster. Solzhenitsyn appreciates the need for humane values and remains one of our most profound moral thinkers.

acknowledgments

I'd like to thank all the taxi drivers who indulged me with discussions on the nature of life in the universe. I have taken the liberty of summarizing some of our conversations in the interests of brevity and quality, but the spirit of these conversations and the central idea raised in each taxi ride have been retained. I thank the team at Harvard University Press, particularly Janice Audet and Emeralde Jensen-Roberts, for their advice and guidance, and Simon Waxman for ideas and suggestions that considerably improved the manuscript. I also thank Antony Topping at Greene and Heaton for representing this work. Finally, I wish to thank my colleagues who, through the years, have helped me develop my interest in and thoughts about life in the universe.

image credits

index